Helium:
A Public Policy Problem

Helium Study Committee

Board on Mineral and Energy Resources

Commission on Natural Resources

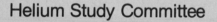

Helium:
A Public Policy Problem

A Report of the
Helium Study Committee
Board on Mineral and Energy Resources
Commission on Natural Resources
National Research Council

NATIONAL ACADEMY OF SCIENCES
Washington, D.C.   1978

NOTICE

The project that is the subject of this report was approved by the Governing Board of the National Research Council whose members are drawn from the Councils of the National Academy of Sciences, the National Academy of Engineering, and the Institute of Medicine. The members of the Committee responsible for the report were chosen for their special competences and with regard for appropriate balance.

This report has been reviewed by a group other than the authors according to procedures approved by a Report Review Committee consisting of members of the National Academy of Sciences, the National Academy of Engineering, and the Institute of Medicine.

This study was supported by Contract No. J0188017 with the U.S. Bureau of Mines, Department of the Interior.

International Standard Book Number 0-309-02742-X

Library of Congress Catalog Card Number 78-52398

Available from

Printing and Publishing Office
National Academy of Sciences
2101 Constitution Avenue
Washington, D.C. 20418

Printed in the United States of America

HELIUM STUDY COMMITTEE

Members	Affiliations
Robert M. Drake, Jr., <u>Chairman</u>	Studebaker-Worthington, Inc. New York, NY
Wiliam D. Carey	American Association for the Advancement of Science Washington, DC
Earl F. Cook	Texas A&M University College Station, TX
Charles W. Howe	University of Colorado (c/o University of Minnesota, St. Paul, MN)
John K. Hulm	Westinghouse Research and Development Center Pittsburgh, PA
Lester B. Lave	Carnegie-Mellon University Pittsburgh, PA
Franklin A. Long	Cornell University Ithaca, NY
F. Clayton Nicholson	c/o Northern Natural Gas Co. Washington, DC
Michael Tinkham	Harvard University Cambridge, MA
Albert E. Utton	University of New Mexico Albuquerque, NM
Irvin L. (Jack) White	University of Oklahoma Norman, OK

Staff

William L. Petrie, <u>Project Officer</u>	Board on Mineral and Energy Resources Commission on Natural Resources

PREFACE

In late summer 1977, the U.S. Bureau of Mines asked the National Research Council's Board on Mineral and Energy Resources for an up-to-date scientific and technical study and an assessment of the long-term needs for helium and the available options and appropriate policy alternatives. The project was accepted.

As a means of expediting the broad orientation of the specially appointed Helium Study Committee to the particulars of its task, a Public Forum was held on November 20-21. Under the general subject, "Helium: Present and Future Needs," the Forum presented overviews and discussion reflecting the major current authoritative opinions and evaluations concerning the problem.

The overviews presented were: <u>Past and Present Uses of Helium</u> by Harold W. Lipper, retired Chief, Division of Helium, U.S. Bureau of Mines; <u>Costs of helium and Present Supplies</u> by Lester B. Lave, Professor and Head, Department of Economics, Carnegie-Mellon University; <u>Future Requirements and Uncertainties</u> by Charles Laverick, Consultant, formerly with Argonne National Laboratory; and <u>Management of Helium Resources</u> by Thomas A. Henrie, Associate Director, Mineral and Materials Research and Development, U.S. Bureau of Mines, and Chairman of the Interagency Helium Committee.

The report has three appendixes: Appendix I is a Bibliography on helium; Appendix II is the proceedings of the November 20-21, 1977, Public Forum on <u>Helium: Present and Future Needs</u>; and Appendix III is a copy of the Invitation to the Forum as well as a number of the responses and recent statements received.

The importance of the prudent management of this natural resource, and the necessity to deal with the problems pertaining to that management without delay, have made the Committee's task an exceptionally challenging one. Circumstance required that that task be undertaken quickly and completed with great speed. We hope that our efforts, reported on here, will make a useful contribution to the solution of the complicated problems related to the management of a critically important resource.

Robert M. Drake, Jr.
<u>Chairman</u>

ACKNOWLEDGEMENTS

Support for this study was provided by the U.S. Bureau of Mines, Department of the Interior, for which the Committee expresses its appreciation. The Committee is grateful for the information and expertise of the speakers, the other participants, and the staff of the Academy Forum.

The willing help of the external reviewers, and others who gave valuable time to help meet the project deadline, was immeasurably important and is immeasurably appreciated.

Finally, and most especially, the Committee expresses its gratitude to one of its members, Dr. Earl F. Cook, who accepted the task of preparing the draft of its report, and carried it through with distinction.

The Helium Study Committee

CONTENTS

FIGURES

TABLES

UNITS

Even though the United States is working toward broader use of metric units, practically all of the published material on helium uses gas-industry terminology, as follows:

Industrial Unit	Definition	Remarks
Mcf	Thousand cubic feet	To convert gas volume units to cubic meters, multiply cubic feet by 2.832×10^{-2}
MMcf	Million cubic feet	
Bcf	Billion cubic feet	
Tcf	Trillion cubic feet	

HELIUM: A PUBLIC POLICY PROBLEM

EXECUTIVE SUMMARY

   Helium is a vital nonrenewable resource needed for
advanced energy and national defense systems, and yet it
is being treated today either as an expendable by-
product of the separation of other materials from some
helium-rich natural gas streams or as a worthless
component of other natural gas being burned for fuel.
Helium is found in air at only 5 parts per million (ppm),
but fortunately there are in the United States almost 100
natural-gas fields containing more than 0.3 percent helium;
such natural gas is considered helium-rich. As far as is
known, 96 percent of the world's helium-rich natural gas is
located in the United States; most of the other 4 percent is
in Canada. It will cost about 800 times as much, in energy
units, to recover helium from air as it does to separate it
from the average helium-rich (.4% He) gas. Helium-rich gas
reserves are being depleted at a rapid rate by the use of
such gas for fuel. Fields now being exploited for natural
gas, representing almost 85% of the measured reserves at the
beginning of 1977, are expected to be essentially exhausted
within 20 to 30 years.

   The U.S. Government initiated a helium recovery and
storage program in 1960 (P.L. 86-777). Somewhat later,
long-term helium purchase contracts were negotiated with
four private companies to separate helium from natural gas
and deliver it to the Bureau of Mines for storage in a
partially depleted natural-gas field. Because government
sales of helium were less than expected, the program proved
to be costly; the government terminated the purchase
contracts in 1973, precipitating litigation that still
continues.

   Because present producing capacity far exceeds current
demand and because the future price is uncertain, there is
little market incentive for conservation of helium by
private producers. In fact, some separated helium is now
being vented to the atmosphere. Most of the helium in
helium-rich gas streams is dissipated with the burning of
the gas as fuel. If present trends continue, there is a
high probability that, early in the next century, when the
currently known helium-rich natural-gas fields will have
been exhausted, helium needs may become great and there will

be little or no helium stored to meet those needs. A policy question arises concerning what should be the national response to this prospect, or to the prospect of a similar status some decades later even if helium is stored now? There are several alternatives:

- Continue Present Policies and Practices

  This alternative could lead to a gradual depletion of the stored helium as it is used to meet federal needs. Once the current litigation is settled, the private companies probably would separate only enough helium to supply current commercial demands; meanwhile, the helium-rich natural-gas fields would be depleted to exhaustion.

- Modify Present Policies to Encourage Helium Separation and Storage by Private Gas Producers

  Modest changes in tax treatment for stored helium and availability of low-cost government-operated storage facilities might encourage private producers to separate and store. A stronger measure would be to levy a "helium tax" on all natural gas that contains over, say, 0.25 percent helium and which is fed unseparated into distribution systems for commercial use. Still other incentive measures are possible.

- Reactivate Government Purchase and Storage Program

  The government purchase and storage program of the 1960s could be reactivated and expanded. Purchase prices could be adjusted to produce selected levels of storage. In this alternative, private industry would continue to build and operate the separation plants. Some market guarantee might be needed in addition to those implied in the purchase contracts.

- Initiate a Government Separation Program

  Another alternative would be for the government to purchase helium-rich natural gas, perhaps in entire fields, build helium separation plants, and store the separated helium. Presumably, this alternative would be resorted to only if it were determined that other alternatives would not yield an adequate amount of stored helium.

2

- **Establish Strategic Helium Reserves with Government Stockpile**

    The present government stockpile of separated helium in storage could be set aside as a strategic reserve, not to be invaded except for reasons of national security; the marketplace would then supply all the helium requirements, including those of federal agencies. Helium-rich gas reserves that are unsuitable as fuel could be acquired and held by a federal agency. Helium-rich fields of natural gas on government land could be withheld from gas production and hence constitute an unseparated reserve.

These alternatives are not mutually exclusive. Which or what mixture of these policy alternatives should be adopted depends strongly on an assessment of the future needs for helium and the expected future costs of obtaining it from the air. Even if major new uses for helium do not develop, it is safe to predict that demand will rise at a rate of around 5 percent per year. Major new uses could arise to greatly extend the need for helium in the next century: superconducting magnets, cryogenic electric transmission lines, and fusion-power applications are examples.

New sources of low-cost helium are conceivable but may remain undiscovered. A more detailed study is needed to improve our understanding of the genesis of helium concentrations in the earth's crust to support more reliable estimates of the resource base. Until a clearer assessment of helium resources is developed, however, there is a strong case for building a substantial government-owned strategic reserve of helium for use in the next century. There is also a good case for encouraging private industry to separate and store helium. How large the reserve should be and in what form is a matter for policy makers to decide. A more detailed study is needed as a basis for these choices. In the meantime, continued atmospheric venting of helium from natural-gas separation plants is contrary to the national interest. The results of this preliminary analysis lead clearly and unambiguously to the conclusion that the venting of separated helium to the atmosphere, either directly or indirectly, should be stopped forthwith. Moreover, because of the massive uncertainties in the resource base, the long-term helium supply, and in the national need, all appropriate measures should be taken in the near term to: seek ways to conserve the helium present in helium-rich natural gas accessible to the helium pipeline and the Cliffside storage reservoir; hold government-owned helium stored in Cliffside as a strategic reserve; and induce or provide incentives for the private sector to store helium while it is readily available.

HELIUM: A PUBLIC POLICY PROBLEM

I. INTRODUCTION

Helium: The Problem

Helium is a vital nonrenewable resource needed for advanced energy and national defense systems, and yet it is being treated today either as an expendable by-product of the separation of other materials from some helium-rich natural gas streams or as a worthless component of other natural gas being burned for fuel. It is being wasted or dissipated into the atmosphere from which it will take about 800 times more energy to recover than it does now from helium-rich natural gas. The prospect of exhaustion of most of the readily recoverable helium resources within the next 20 to 30 years poses a clear-cut issue of the national interest, because there appears to be no substitute for helium for certain uses that may become very important early in the coming century. Obscured from public view by current dilemmas associated with energy supply and an unstable economy, mired in technological and legal uncertainties, public policy for management of the nation's helium resources appears inadequate and inflexible. The helium question has been addressed in two published reports of the National Research Council (Committee on Resources and Man, 1969, pp. 10-11; Physics Survey Committee, 1972, pp. 42-43) and, at much greater length, in ERDA-13 (Energy Research and Development Administration, 1975).

Helium: The Resource

Helium is a light, inert gas, unique in its properties and critical uses. It becomes a liquid when cooled below $4.27°K$ (Kelvin, the equivalent to Celsius degrees above absolute zero) and, most importantly, unlike other gases, helium remains liquid to the lowest temperature yet achieved, almost absolute zero ($-273.16°C$). This latter property makes it indispensable as a refrigerant to cool certain substances to temperatures at which these substances

4

become superconductive, that is, have vanishingly small electrical resistance. Furthermore, at about 2 degrees above absolute zero, liquid helium becomes nearly a perfect heat conductor.

The earth's atmosphere contains an equilibrium concentration of 5 parts per million (ppm) helium. However, natural gas contains much greater concentrations of helium, even as high as 99,900 ppm (nearly 10 percent) in USBM Sample No. 11689 from the Permian Coconino Formation in Apache County, Arizona. Helium-rich natural gas is defined as that containing 0.3 percent (3,000 ppm) or more helium. Most natural gas contains helium in concentrations of 0.1 percent or less. World reserves of helium-rich natural gas are overwhelmingly concentrated in the United States, principally in one area of the southern Great Plains (Figure 1). Outside the United States, the only nation with natural gas containing 0.3 percent or more helium is Canada, whose helium-rich gas was calculated to contain 4.3 Bcf helium (Energy Research and Development Administration, 1975, Table X, p. 76). As used by the U.S. Bureau of Mines (Miller and Tipton, 1977), helium resources as of January 7, 1977, included [in billions of cubic feet (Bcf)]: stored helium 38.5, measured (proved) reserves 97.3, indicated (probable) reserves 49.6, measured paramarginal and submarginal reserves 66.4, indicated paramarginal reserves 6.7. In the Miller and Tipton report, the term resources and reserves are used interchangeably and the 5 quantitative estimates, representing categories covering a wide range of uncertainty, are added together to produce a total of 258 Bcf of "identified helium resources." Measured and indicated reserves refer to helium in helium-rich natural gas. Paramarginal reserves refer to helium in natural gas containing at least 0.1 percent and less than 0.3 percent helium. Submarginal reserves refer to helium in natural gas contining less than 0.1 percent helium. As used herein, reserves refer solely to measured (proved) reserves of helium in helium-rich natural gas.

Two thirds of the helium contained in U.S. natural gas reserves is in the helium-rich fields, of which there are about 100 in 10 states. Only one third of the helium is in the much more abundant helium-lean gas reserves. Proved or measured U.S. helium reserves (in helium-rich natural gas, including shut-in fields, but not including stored helium) have been dropping since the start of 1964 (Figure 2) at an average rate of 7 Bcf a year. During this same period, helium "production" from the natural gas withdrawn from helium-rich fields averaged about 8 Bcf a year. Of the approximately 105

HELIUM RESOURCES OF THE U.S.

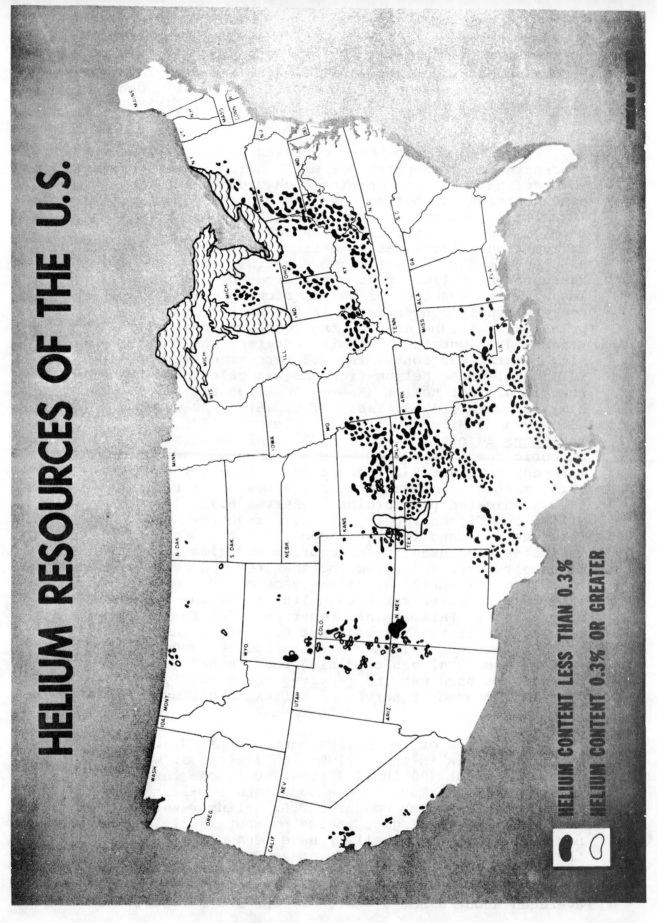

HELIUM CONTENT LESS THAN 0.3%

HELIUM CONTENT 0.3% OR GREATER

Figure 1. Helium resources of the U.S.

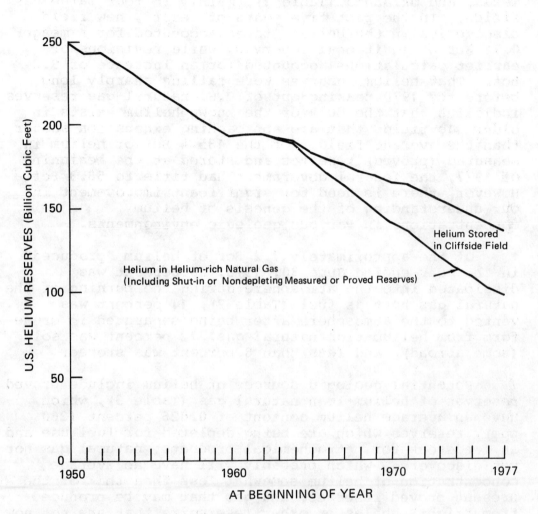

Figure 2. U.S. helium reserves, 1950-1977

Bcf of helium produced during this 13-year period, 38 Bcf were stored in the Cliffside gas field in the panhandle of Texas, mainly under a conservation program managed by the U.S. Bureau of Mines.

Eighty-six percent of the helium reserves of the United States at the beginning of 1977 were in Kansas, Texas, and Oklahoma (Table 1), mainly in four large gas fields. In the past five years of record, new field discoveries in the United States accounted for a meager 0.71 Bcf of additional reserves, while revisions of earlier calculations accounted for an increase of 9.2 Bcf. That helium reserves were falling sharply long before the 1970 peaking-out of U.S. natural-gas reserves indicates that the bulk of the known helium exists in older gas fields that are approaching exhaustion sooner than the average field. Of the 135.4 Bcf of helium in measured (proved) reserves and stored at the beginning of 1977, the federal government had title to 58.3 Bcf. However, there is need for significant improvement in our understanding of the genesis of helium concentrations in various geologic environments.

Of the approximately 7.2 Bcf of helium "produced" in the year ending June 30, 1976, 66 percent was dissipated into the atmosphere during the burning of the natural gas host as fuel (Table 2), 14 percent was vented to the atmosphere after being separated in crude form from helium-rich natural gas, 12 percent was sold (some abroad), and less than 8 percent was stored.

Potential geologic sources of helium include proved reserves of helium-lean natural gas (Table 3), which have an average helium content of 0.026 percent (260 ppm), reserves which are being depleted for fuel use and as feedstock for petrochemical products; natural gas not yet discovered, which probably will have an average concentration of helium somewhat less than that of the present proved reserves; and gas that may be produced from "tight" shales or other reservoirs that are not now economic sources. The helium content of the few samples of natural gas from Devonian shales of the Appalachian and Michigan Basins so far analyzed by the U.S. Bureau of Mines is less than 0.1 percent (B.J. Moore, USBM, Oral Communication, December 1977). The helium content of the geopressured fluids (under high pressure caused by great depth) in deep Gulf Coast strata is apt to be less than 0.1 percent because the natural gas in known reserves of that region has a very low helium content. In 1975 an Energy Research and Development Administration projection (Figure 3) indicated a rapid decline after 1980 in the helium content of helium-rich

Table 1: U.S. Helium Reserves* at Year's Beginning (in Bcf)

STATE	1973	1974	1975	1976	1977	1973-77 change
Kansas	55.34	51.58	48.05	44.63	45.41[1]	-17.9%
Texas	39.67	36.14	32.70	29.45	26.33	-33.6%
Oklahoma	13.35	12.31	13.48[1]	12.43	11.50	-13.9%
Wyoming	3.26	6.28[1]	6.28	6.32[2]	6.31	+93.6%
Utah	3.03	3.91[1]	4.10	4.14[3]	4.13	+36.3%
Colorado	1.66	0.44[1]	0.76[4]	1.07[5]	1.02	-38.9%
Arizona	1.79	1.73	1.69	1.66	0.82[1]	-54.2%
New Mexico	0.73	0.38	0.60[1]	0.75[1]	0.63[1]	-13.7%
Montana	0.58	0.58	0.58	0.58	0.47[1]	-19.0%
West Virginia	0.11	0.12	0.12	0.12	0.12	+ 9.1%
	119.52	113.47	108.36	101.15	96.75	-19.1%

Source: R. D. Munnerlyn, USBM Division of Helium, personal communication, December 9, 1977.

1 Change from previous year resulted from revision of earlier reserve estimates.

2 Includes a new field - Baxter Basin South in Sweetwater County.

3 Includes a new field - Boundary Buttes North in San Juan County.

4 Includes a new field - Doe Canyon in Dolores County.

5 Includes a new field - South Canyon in Garfield County.

* Proved or measured reserves; in addition, at the beginning of 1977, the Bureau of Mines estimated about 51 Bcf in indicated (probable) reserves.

Table 2. Disposition of Helium "Produced"
in Year Ending June 30, 1977 (Bcf)

Separated and sold	0.88	12%
Separated and stored	0.54	8%
Separated and vented	0.98	14%
Dissipated in fuel gas	4.74	66%
TOTAL	7.14 ±	100%

Source: Morgan, 1977.

Table 3. Helium Contained in Helium-Lean Natural Gas
(as of Beginning of 1973) in States where Helium-
Lean Gas Averages over 0.1 Percent Helium

STATE*	Average helium content of helium-lean gas (%)	Contained helium (Bcf)
Texas RRC 7B	0.153	1.02
Ohio	0.147	1.69
Kansas	0.145	0.42
Kentucky	0.125	1.17
Illinois	0.123	0.67
Michigan	0.120	1.56
Texas RRC 9	0.119	1.86
Indiana	0.117	0.10
Texas RRC 10	0.106	2.88
Average	0.123	Subtotal 11.37
Other states and Texas districts	0.022	51.43
Average	0.026	Total 62.80

* In Texas, Railroad Commission (RRC) districts

Source: Moore, 1976, Table 3, p. 7.

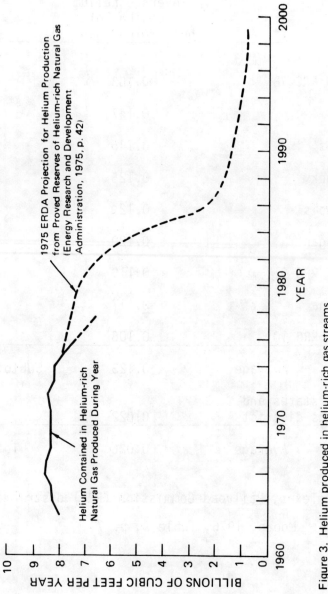

Figure 3. Helium produced in helium-rich gas streams

natural gas produced. Actual production for the year 1976 suggests that this decline has already started.

The estimates are based on different assumptions. However, if we use recent estimates (Exxon, 1976) of natural gas remaining to be discovered and produced in conventional reservoirs in the United States, we can make some calculations of the helium potential in undiscovered gas (Table 4). Calculated in this way, the total helium content of natural gas remaining to be discovered and produced is less than a third of the 568 Bcf estimate of the Bureau of Mines (Morgan, 1977) for the total of "indicated reserves" and "undiscovered resources." Even this smaller estimated amount may be unduly optimistic, since the calculations assume that helium-rich gas will be discovered at a rate proportional to its present quantitative relation to helium-lean gas, contrary to the experience of the past 20 years. If the U.S. Geological Survey estimates (Miller et al., 1975) are used for the calculations of Table 4, the helium total almost doubles (315 Bcf).

The energy cost of separating helium (only) from a mixture of other gases is inversely proportional to the helium content of the mixture (Figure 4). To illustrate, producing a cubic foot of helium from air requires processing 800 times as much gas (in this case, air) as would be required to produce a cubic foot of helium from natural gas containing 0.4 percent helium, which is the average of the gas that has been flowing through the mid-continent separation plants. Because the principal cost in a separation plant is the cost of the energy needed to liquefy the components, the pecuniary cost is closely proportinal to the energy cost.

Present and Future Uses of Helium

At present the major uses of helium are in cryogenic applications -- for purging and pressuring, in welding, and for controlled atmospheres and breathing mixtures (Figure 5). Minor amounts are used in leak detection, for lifting and heat transfer, and in chromatography. Federal agencies consume slightly more than half the marketed production. Until 1962, the U.S. Bureau of Mines was virtually the sole source of helium (Figure 6). Since 1969, the private sector has supplied more than half the helium marketed.

Demand for helium in the future on a scale larger than at present will depend on the development and deployment of sophisticated technologies that either do

Table 4. Helium Content of Potential Gas Reserves in the United States*

	Potential natural gas (Bcf)		Probable helium concentrations**	Probable helium content (Bcf)	
	Exxon	USGS		Exxon	USGS
Offshore	101,000	107,000	0.007% [1]	7	7
Inland	186,000	377,000	0.068% [2]	126	256
Probable growth of known fields	111,000	201,600	0.026% [3]	29	52
TOTALS	398,000	685,600		162	315

[1] Average helium content of natural gas being produced from coastal zones of Texas and Louisiana and in all of California.

[2] Average helium content of all measured natural-gas reserves in the U.S., including helium-rich gas.

[3] Average helium content of helium-lean natural gas measured reserves in the United States, most of which are inland.

* Based on estimates made in 1975 (a) by Exxon USA and (b) by U.S. Geological Survey (Miller et al., 1975)

** Based on Moore, 1976, Table 3, p. 7.

14

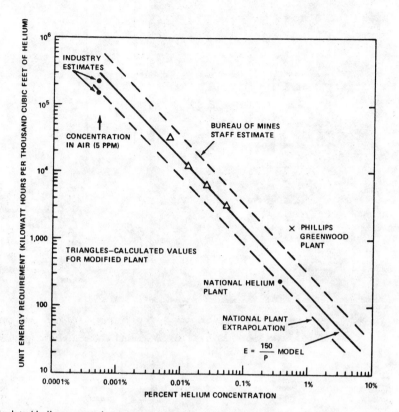

Actual and calculated helium extraction energy requirements in kWh per thousand cubic feet of helium vs. percent helium concentration. Solid line corresponds to the simple model E = 150/P described in ERDA-13. Dashed lines correspond to upper and lower limits to energy estimates for conservation-type plants.

Source: Energy Research and Development Administration, 1975, p. 66.

Figure 4. Energy requirements to extract helium

U.S. HELIUM CONSUMPTION

601 million cubic feet

USES

OTHER	20%
CONTROLLED ATMOSPHERES AND BREATHING MIXTURES	12%
WELDING	16%
PURGING AND PRESSURIZING	18%
CRYOGENICS	34%

USERS

U.S. COMMERCIAL MARKET 48%

OTHER
ERDA
DOD
NASA

FEDERAL MARKET 52%

-100%-
-80-
-60-
-40-
-20-

BASED ON 1975 MRI SURVEY

BUREAU OF MINES

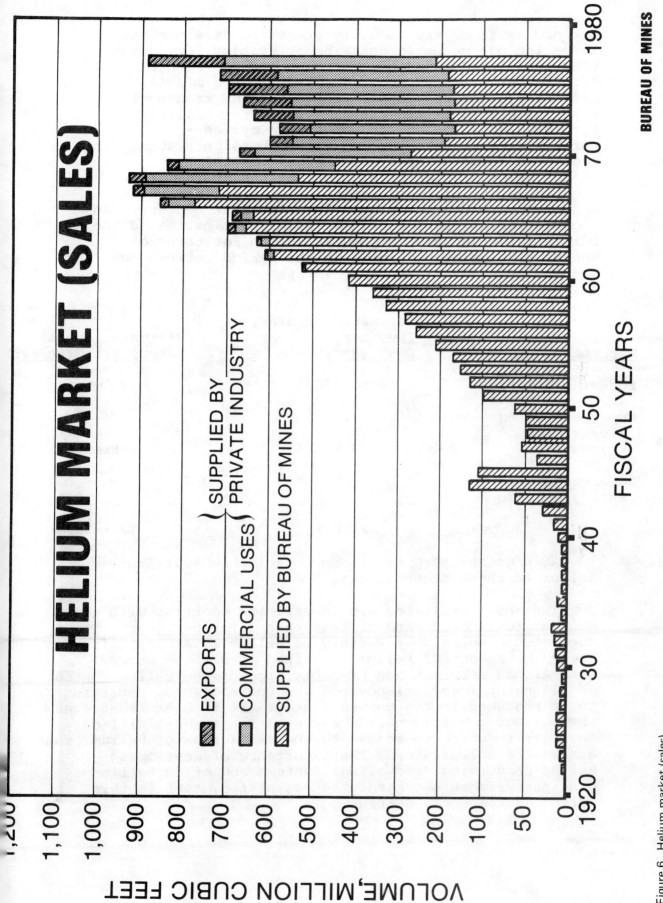

Figure 6. Helium market (sales)

not now exist or are in early stages of development.
These include magnetic containment systems for fusion
reactors, breeder and high-temperature gas reactors,
high-temperature gas turbines, laser-based missile-
defense systems, propulsion units for new transport
systems, helium refrigeration systems for military
aircraft, advanced energy conversion cycles--
particularly magnetohydrodynamic conversion systems--and
superconducting devices for energy conversion,
transmission, and storage. None of the helium demands
created by any of these technologies will likely
approach maturity until well into the 21st century. An
Argonne-NSF Advisory Committee study has estimated the
cumulative inventory demand for helium for three of
these emerging technologies in the period between now
and the year 2050 (Laverick, 1975):

Technology	Estimated Cumulative Inventory to year 2050 (Bcf)	Estimated Annual Makeup Rate (Bcf/yr)
Fusion reactors	50-100	5-10
Superconducting electricity transmission	425	not estimated
Superconducting energy storage	54	5.4
Totals	529-579	10.4-15.4

At present there exists no conceivable substitute for
helium in these technologies.

The above estimates are in striking contrast with the
present level of demand of less than 1 Bcf per year. The
suggested cumulative inventory of helium is at least four
times the amount of helium contained in all the proved
natural gas reserves and probably exceeds the helium content
of all natural gas, discovered and undiscovered, remaining
to be produced in the United States. It is clear that even
the maximum possible helium-conservation program will only
buy time in which to adjust to the extraction of helium from
air as the sole source. The importance of developing
helium-conserving (recycling) systems and of curtailing
dissipative uses can hardly be overstated, even in this
period of temporary surplus.

The Helium Conservation Program

This section is based on the summary description of
the federal government's helium program contained in the
statement by Morgan (1977, pp. 2-3.)

Although the federal government's interest in
helium production began as early as 1917, it was the
Helium Act of 1925 that established the first helium
production and sales program (Morgan, 1977). This Act
gave jurisdiction to the Bureau of Mines to produce and
sell helium to government agencies. The 1925 Act was
amended in 1937 to permit the sale of helium not
required to meet government needs to private users and
to provide that the costs of helium production be
financed by revenues accruing from helium sales.

In 1960, an interagency study found that the
existing Bureau of Mines program would not result in the
production of enough helium to meet the demand that was
estimated to occur after 1985. This concern led to an
extensive amendment of the 1937 Act. In the 1960
revision (P.L. 86-777), the Bureau of Mines production
was to be supplemented by helium purchased under
contracts with private producers. The costs of the
program were to be financed out of income from helium
sales. Until that income became adequate to meet
program costs, the Bureau of Mines was authorized to
borrow operating funds from the U.S. Treasury. The
borrowed funds and accrued interest were to be repaid
within 25 years; however, the Secretary of the Interior,
under the Act, could extend the repayment period by up
to 10 years.

In implementing the provisions of the 1960 Helium
Act, the Bureau of Mines entered into purchase contracts
for helium with four private companies: Phillips
Petroleum, Cities Service Helex, Northern Helex, and
National Helium. The Bureau contends that the
production from the private companies, together with the
Bureau's own production, was intended only to produce
helium to meet current needs and to create a government-
owned stockpile of 41.5 Bcf to be stored in the
partially depleted Cliffside natural-gas field near
Amarillo, Texas. Furthermore, this stockpile was
intended only to meet essential government needs after
1985. By 1967, the demand for helium by federal
agencies, excluding contractors, began to decline such
that the combined production of the Bureau, its contract
producers, and those private producers attracted to the
helium production business during the period of high
demand substantially exceeded demand. The decline in
federal demand, the alleged, but still unverified,

discovery of substantial new helium reserves, and the
development of improved helium production technology
were the reasons given by the Secretary of the Interior
for terminating the helium purchase contracts in 1973.
(Cancellation of the contracts on these grounds has
been challenged by the private parties to the contracts.
The litigation of these cases is described in Section II
below.) Subsequently, helium has been extracted at
multiproduct, private plants and vented to the
atmosphere, initially at a rate of more than 2 Bcf per
year, and until December 1977, at a rate of about 1 Bcf
per year. The reduced venting rate results from plant
shutdowns and a Bureau of Mines storage policy initiated
in 1975, that permitted private producers to store their
helium in the Cliffside reservoir. At present,
approximately 39 Bcf of helium are in storage in this
reservoir. Another 3.7 Bcf is contained in the natural
gas that remains in the field.   The Bureau maintains
that no more helium need be stored to meet the essential
needs of federal agencies, at least until the early
years of the 21st century (Morgan, 1977, p. 3).

Elements of the Policy Problem

The public policy predicament is complex. The
Executive Branch of the federal government interprets
its principal responsibility under the Helium Act of
1960 to be that of providing for the future needs for
helium by federal agencies. According to a survey of
government users, helium now in storage and belonging
to the government is ample to meet the currently
anticipated needs of the agencies into the coming
century (Morgan, 1977). Helium being extracted by the
U.S. Bureau of Mines is more than sufficient to meet
current demand for helium by government agencies. That
helium being separated from natural gas by private
companies more than meets the present needs of the
private sector, at a price well below that established
by the Bureau of Mines. Thus, there is no sense of
crisis as an incentive to generate a conservation policy
for helium in the short term.

The case for helium conservation is long-term,
based on the interest of future generations in
preserving a supply of helium for technologic options at
the expense of present generations. A policy decision
to capture and store helium while it is still possible
to do it cheaply, for use in a future when it may be
very much more expensive, would be a prudent investment.
 The policy alternative to conservation is to ignore
contingencies of the future and thus, perhaps, impose a
heavy resource constraint in the future. From a

geological viewpoint, the conservation policy is reversible; the alternative is irreversible.

The management policy for helium is complicated, as well, by past decisions. For a decade, beginning in the early 1960s, the federal government initiated and managed a program designed to accumulate stores of helium sufficient to meet these needs expected for the space and national defense programs. However, changing program requirements reduced the projected demand for helium to a level so low that by the early 1970s the Secretary of the Interior terminated the purchase contracts for helium produced by private firms. Vigorous litigation ensued and continues. The issues of a rational policy for management of helium are confused because of the uncertain outcome of several lawsuits. While this adversary process works itself out, most of the helium being produced from helium-rich natural gas is being permitted to escape into the atmosphere.

As if these problems were not enough, policy making is perturbed by the question of appropriate government-industry relations and respective roles relative to helium conservation and storage. A stated objective of the Helium Act of 1960 was the encouragement of helium conservation by private industry. Yet that same Act requires the government to sell helium from its stockpile to anyone willing to pay a price that will allow recovery, with compound interest, of the cost of the storage program. Not only has no special economic incentive, other than a low storage fee, been made available to industry, but the market effect of the government's existing stock of stored helium inhibits industry's storage of helium in anticipation of a distant future profit. Furthermore, it is not inconceivable that helium could be re-declared a strategic material at some time in the future, its consumptive uses curtailed by law, and the market price accordingly affected adversely. For industry to be an important factor in helium conservation such that government will not have to bear all the costs of conservation, a broad range of questions concerning long-term stability and conservation incentives in public policy remains to be faced and resolved.

Finally, the problem of choice is affected by perceptions of strategy for helium relative to national energy policies and technological trends. Helium can be viewed, on the one hand, as a free-standing issue of resource conservation. On the other hand, it can be seen as an important element in energy management. Research, development, and investment in fusion and laser-fusion technologies, magnetohydrodynamic power production, and all superconducting devices are dependent upon helium as a refrigerant to attain the very low temperatures that permit materials to become superconductors. From this perspective, national energy policies and helium-conservation policies are interdependent. Government support of research and development of new energy technologies that require, or can make use of helium, may total $6 billion over the next few years. The cost of conserving helium might be regarded as an insurance premium to protect that large research and development investment, which in itself may understate the present value of the plan that will depend on superconductivity.

This outline of the complexities of the policy-making problem is relevant to the question of whether or not helium should be conserved. Scientific, technological, and social assessments of the importance of helium must be thought about along with considerations of resource management, economics, and institutional arrangements. The issue has been pursued in these aspects.

The considerations began with the following consensus perspectives:

- Helium is an essential resource for which, in some uses, there is no substitute, and in other uses, no suitable substitute.

- Future demand for helium will be substantial and perhaps very large in comparison with present demand. At current rates of depletion, helium-rich natural gas will be virtually exhausted in 20 to 30 years. Helium-lean natural-gas reserves, although much more abundant, cannot be counted upon as a large potential source of helium much beyond the exhaustion of the helium-rich gas (see Table 3).

- Public policy on helium conservation should no longer be based solely on federal government requirements, but on national needs.

- Foreclosure of the availability of cheap helium after the end of the present century engenders long-term risks and uncertainties, especially as related to emerging energy technologies.

II. DIMENSIONS OF THE HELIUM PROBLEM

The National Interest in Helium Conservation

A New Context for Conservation

The perception of the national interest in helium conservation appears to have changed since 1960 when the existing helium conservation structure was set up. Then, the concern was more immediate and related to anticipated needs in space exploration and defense technology. Today, the concern is longer range and related to energy needs and expected technological options at the beginning of the next century. Moreover, today there is a broader perception of the rapidity and inevitability of the depletion and coming exhaustion of the nation's natural-gas resources, which constitute the only large inexpensive source of helium. The relatively short-term payback scheme organized under the Helium Act of 1960 may have fitted the then existing circumstances and perspectives, but it does not fit today's changed circumstances and new perspectives.

The Question of a Strategic Reserve

Natural gas is used extensively to heat our homes and to provide fuel for industry. It is difficult to reserve helium-rich natural gas until such time as the helium might be worth extracting on a market-demand basis. If the natural gas is burned before the helium is removed and the helium is dissipated to the atmosphere, some 1,000 times more energy is required for recovery than that necessary to extract it from natural gas (containing 0.3 to 0.5 percent helium) before combustion.

If the nation is to have an adequate reserve of helium for use in the energy and defense technologies at the beginning of the 21st century, we have the option to pay the costs of additional helium extraction and storage today or face the strikingly higher costs of extracting helium from its ultimate reservoir, the air. Furthermore, it should be noted that the difference in direct energy requirements is not a full measure of the cost difference to society, because the cost of energy itself, 50 years from now, may be several times what it is today.

The existing helium stockpile cannot be regarded as a strategic reserve that will ensure the availability of technologic options 30 to 50 years hence. First, it may be too small. But of greater significance is the fact

that the Bureau of Mines is required by the 1960 Helium
Act amendment to sell from the present stockpile to
anybody willing to pay the government's price necessary
to recoup the costs of helium separation and storage.
Under this scheme, there is no guarantee that any helium
will remain for future technology, most of which would
not involve much dissipative use of helium, but could
recycle it indefinitely. It is conceivable that
dissipative uses (welding, breathing mixtures, etc.)
could, if permitted, exhaust the present stockpile
before the inventory demand of the new technologies
develops. Thus, consideration might be given to the
establishment of a strategic reserve of helium, to
replace or supplement the present stockpile. To convert
the stockpile to a strategic reserve from which sales
could be made only with Congressional authorization or
review not only would protect the future better but
would greatly reduce one of the elements of uncertainty
that deter private industry from storing helium for
future sale. Before discussing this question further,
let us turn to a more general review of the problems
that plague decisions of this nature.

Management of a Finite Natural Resource

In attempting to get the maximum social benefit
from a finite natural resource, we must ask: (1) What
should be the shift in the time pattern of consumption?
(2) For any given amount of consumption reduction, when
and how should the conserved resources be employed? (3)
What would be the cost of saving for the future all the
helium in present helium-rich natural gas streams? The
key to the answers is an assessment of (a) the future
benefits to be obtained from future consumption versus
current use; (b) the opportunity costs of deferral; and
(c) the magnitude of the finite resource, which may
change with accidents of invention and acts of nature or
with the abilities and desires of society.

Some experts question the desirability of
undertaking a monetary analysis, while others believe
that estimation of future benefits calls for social and
technologic assessments as well as economic projections.
The opportunity costs of consumption deferral may be
calculated in dollars, but will entail nonmarket or
social costs as well. Finally, the estimation of the
magnitude of a finite geological resource, such as
copper, uranium, or natural gas and its contained helium
is complicated by the uncertainties of geology,
technology, the future value of energy, as well as the
political constraints.

The marketplace may help to answer both questions posed above, but the naive projection of recent economic history appears increasingly hazardous. One economic approach would be to reason as follows: If one had had a billion dollars in 1900 and had been intent on improving the economic health of America (or merely wished to augment the initial bankroll), should one have invested in natural resources (such as rich iron or copper deposits), in industrial companies, in research and development, or in public health services and education? Even in hindsight, the choice is not obvious.

Historical statistics show that the prices of iron and copper fell (in constant dollars) between 1870 and 1970. Therefore, specific investments in natural resources may not have paid off well, although it could be argued that the massive increases in production and utilization of these two metals added greatly to the physical plant and economic vigor of the nation. For many mining companies, the increase of ore volume and the advancement of technology more than compensated for the lower prices of the contained metal. Over this same period of 100 years, investments in industrial corporations produced a handsome return, perhaps 7 percent per year in constant dollars. However, the fact that industry could not have existed, let alone prospered, without large quantities of cheap natural resources, including inexpensive energy, cannot be overlooked. Returns on investments in research and development or in improved health and education between 1870 and 1970 appear to have been substantial. In summary, it could be argued that, in retrospect perhaps, investment of national savings in the development and exploitation of finite natural resources over the past hundred years may have been in some cases less wise than investment in other economic activities.

For the future, it could be maintained that an investment of national savings in conservation (deferred consumption) of a natural resource should meet the test of returning at least as much to the economy as would an equivalent, additional investment in any other productive sector. Under this criterion, the test would require that the expected discounted future real value of helium equal or exceed the cost of isolation and storage discounted at the anticipated return from the best available alternate investment of the savings that might be diverted to helium conservation. To an economist, then, it would appear that, if a higher pecuniary return can be expected from investment of public or private savings in other ways than in conserving helium, we should neither conserve helium nor be concerned about its dissipation into the atmosphere. If, on the other hand, a higher rate of return can be anticipated from investment in helium conservation, then public funds should be diverted to this use.

There are, however, some who disagree with this approach to making a decision about the conservation of a finite, exhaustible resource, and who would question the use of a discount rate based on present opportunities for pecuniary profit to determine the implied cost of preserving options for future generations.

To separate and store all of the helium from present helium-rich gas streams should cost less than $100 million a year, an amount equivalent to one-tenth of a cent for each gallon of gasoline sold in the United States in a year. That cost will diminish as the natural gas is depleted. The cost of recovering and storing all the helium in helium-rich natural gas (about 90 Bcf) would be less than $1 billion, equivalent to one cent on each gallon of gasoline sold in the United States in one year. In other words, the cost to the government or to each individual, in terms of present opportunities foregone, would be very small. Then, the basis for the decision becomes the value we put on the assurance of having a helium reserve and in choosing among alternate gifts such as improved environment, better health care, or more basic scientific information.

The nature of the supply of a finite natural resource must be clear before an effective management program can be designed. "Finite supply" might mean a fixed stock of uniform grade of concentration with well defined boundaries, unextendable through further exploration. The supply of mercury in the United States appears to be such. Many resources exhibit a spectrum of grades with the volume available increasing at lower grades. Iron and aluminum are examples. In such cases, supplies are almost indefinitely extendable, but at increasing energy and environmental costs.

The demand for the resource is important in managing it. If no substitutes are available and the goods made from the resource base have no substitutes, demand will be inelastic and price will be sensitive to the quantity available.

When a resource is difficult to expand, or when demands are inelastic, today's use or dissipation of the resource will have important implications for future uses, provided, of course, that the demand for that resource continues in the future. Future users may be forced to do without or to incur very high costs. It is particularly important under such conditions to give explicit consideration to alternative management options, explicating the implications of each as fully as possible.

Policy-Making under Conditions of Uncertainty

Society generally is willing to pay for insurance to avoid certain types of uncertainty. Insofar as our policy decisions today increase uncertainties to be faced by future generations, we create a cost to be borne by them. This must be considered in deciding the equity issues between present and future.

In the case of helium, the costs of supply from the various sources and the associated quantities, while not known with certainty, are reasonably well established. The large uncertainties appear on the demand side, primarily in the questions of the timing and growth rates of: (a) fusion power, (b) development of devices using superconducting magnets, and (c) development of cryogenic transmission systems, and secondarily in the possibility of development of substitute technologies that do not need helium.

In relation to the storage of helium, two kinds of decision error could be made: (a) to store helium, only to find no future uses for it; or (b) not to store and later to find that certain potentially highly beneficial uses are precluded by the high costs of helium. If these possible outcomes were equally likely, and if the social losses occurring in each case were of the same magnitude, it would be foolish to pay for separation. However, most people are risk averse i.e., are willing to pay more than the actuarial value of insurance in order to reduce uncertainty. It is essentially impossible to forecast the future demand for helium. A high demand would warrant current storage; a low demand would make this investment seem foolish in comparison with other possible investments. With uncertainty so great, it is impossible to advocate either storage or non-storage with certainty.

No one knows the future price of helium or the timing of price changes. The price could rise to much more than $3,000 per thousand cubic feet (Mcf) if energy prices increase and technological change is less than expected. Indeed, the price could rise to more than $3,000 per Mcf as early as the year 2000 because of new technologies that use a great deal of helium. On the other hand, the helium-consuming new technologies may never mature, and the demand for helium would then remain small. This uncertainty tends to scare off investors in storage, particularly when a period as long as 50 years may be required to recoup the investment.

Current Issues

Venting

The 1973 termination of the helium supply contracts between the Bureau of Mines and the four suppliers (Northern Helex, Cities Service Helex, National Helium, and Phillips Petroleum) caused an immediate and unfortunate national loss of 3.4 Bcf per year. Because the suppliers who could not shut down and had no means of storing the helium gas, it was allowed to escape at the rate of approximately 2.2 Bcf per year, either by direct venting to air or by return to the natural gas stream for dissipation in the ultimate combustion of the latter.

The wastage of this natural resource, which has a value of perhaps $50 million annually if it could be sold at current market prices, is the question at issue. Venting was caused by the contract cancellation resulting law suits and the lack of suitable storage space available to the companies. Whatever the cause, a modicum of planning could have averted it or stopped it long before now.

In 1975, the Bureau of Mines attempted to mitigate venting by allowing the suppliers to store helium on their own behalf in the Cliffside field, subject to payment of certain storage costs. Northern Helex took advantage of this offer in September 1975, and since then has stored helium at the rate of about 0.6 Bcf per year. In 1976, Phillips Petroleum shut down its plant at Dumas, Texas, reducing the helium supply by about 0.5 Bcf per year. Phillips continues to produce about 0.4 Bcf per year at Sherman, Texas, most of which, beginning in December 1977, is delivered to the government for storage. The production of about 0.6 Bcf of helium per year extracted by Cities Service at Jayhawk, Kansas, constitutes the present venting problem.

Although venting has declined from 2.2 Bcf to less than 1 Bcf per year, wastage is still significant. Supplier representatives at the recent National Academy of Science Public Forum on helium suggested several factors contributing to their reluctance to store in Cliffside:

- Lack of a free market because of the government's ability to influence the future price of helium

- Federal tax policies under which helium stored cannot be expensed until it is finally sold

- Potential <u>ad</u> <u>valorem</u> taxes imposed by states and local communities on stored helium

- Uncertainty of the outcome of litigation on helium royalty payments to landowners and gas suppliers

- Limited duration of the Bureau of Mines storage contracts.

Financing

The original financial plan of the Bureau of Mines Helium Conservation Program called for helium purchases to be financed by loans from the U.S. Treasury. It was anticipated that, by 1971, income from the sale of helium not only would exceed the interest on the loan from the Treasury but also would allow commencement of repayment of the debt, which was supposed to be paid off completely after 35 years.

This plan failed. In 1971, Bureau of Mines sales of helium were a factor of 6 lower than originally projected, because of two developments; first, the total market for helium was about half the original projection, and second, private suppliers found it advantageous to offer helium at $25 per Mcf in contrast to the Bureau's price of $35 per Mcf, thereby capturing about two thirds of the market.

According to data supplied by Dr. John Morgan, Acting Director of the Bureau of Mines, the program indebtedness on July 1, 1976, was $412 million. In the following year, the interest on this debt was about $24 million, whereas the net income from sales of helium was only $1 million. It is clear that, even without further purchases of helium for conservation, the net income from sales is at present far short of covering the interest on the debt.

Much has been made of this financial problem in high government circles. Indeed, the cancellation of the helium supply contracts in 1971-1973 can be attributed primarily to this issue. However, the actual balance sheet shows that, against its $400 million

investment, the government possesses within the
Cliffside storage field about 40 Bcf of helium, an asset
that could generate incomes in excess of $1 billion if
it could be sold at current market prices. Under these
circumstances, the conservation program could prove to
be a prudent investment on financial grounds, not
counting its probable vital role in future U.S.
technology.

Litigation

The Helium Conservation Program has resulted,
directly and indirectly, in a series of complex lawsuits
that to a large extent, are still before the courts.
These suits fall into two general classes, those
concerned with the value of helium as extracted from the
earth and those resulting from the termination by the
Bureau of Mines of its contracts with suppliers in 1973.

With respect to the original-value suits, it
appears that, when contracts were first negotiated with
landowners and gas producers for helium-rich natural
gas, no specific value was assigned to the helium
itself, since, unlike the hydrocarbons, helium has no
value as a combustible material. However, the Helium
Act and the resulting conservation program established
helium as a valuable resource. The landowners and
producers of helium-rich natural gas claim ownership of
the helium and are seeking royalty payments from helium
extractors in a series of lawsuits. These legal actions
have substantially added to the business risk and
exposure of the various helium extractors.

In these lawsuits the U.S. Government is involved
as a third party in several instances. The government
could be liable under the helium-purchase contracts to
which it is a party. These contain provisions that may
require the government to indemnify the helium-
extraction companies for any amount the extraction
companies may be required to pay the landowners and
producers in excess of $3 per Mcf of helium.

These uncertainties as to ownership, cost
expectations, and indemnification liability constitute
substantial impediments to the conservation program.
Helium extracted and placed in storage carries with it
the risk of substantial but unknown future payments to
landowners and producers.

A second general class of lawsuits resulted from
the attempt by the Bureau of Mines to terminate helium
purchase contracts, commencing around 1971 and becoming

effective in 1973. The government attempted to terminate its contracts up to 12 years before the scheduled termination dates. This termination action and subsequent litigation have contributed another substantial element of uncertainty in the helium storage program.

Stockpile-Management Policies

From the evidence available to this committee, the management of the helium program by the Bureau of Mines seems to have been excellent from several viewpoints. We note that superior technical management was exercised in commissioning and operating the Cliffside storage field and its associated gas-collection system, which is unique in the world. Beyond this, the achievement of a stored reserve of 39 Bcf of helium certainly impressive. However, it seems important to examine the causes of the failure of the conservation program to break even financially by 1971.

The basic difficulty is that we are dealing with a resource in which at present there is a surplus, since the quantity of helium separated from natural gas bound for market is greater than the quantity currently demanded. This helium resource is coupled with the U.S. supply of natural gas from certain regions. The natural gas will be consumed in the next two or three decades to supply U.S. energy needs. If helium is not extracted from helium-rich natural gas while such gas is still available, the end of the natural-gas reserves will also signal the end of U.S. helium reserves. In other words, the present helium surplus will be succeeded by a shortage.

If the Bureau of Mines is to be criticized for its role in the conservation program, it should be on the grounds of attempting to sell helium from the stockpile into a surplus market. If its quasimonopoly on intragovernmental sales were removed, the Bureau's market at $35 per Mcf would collapse, as other government agencies move to buy lower-priced helium from private suppliers.

There may be no pricing strategy that the Bureau of Mines can pursue that will recoup the investment in the near term. The original design based on break-even financing by 1971, proved to be in error. As one option, the Cliffside reserve might be kept intact until the end of the surplus period, which is estimated to be in the mid-1980s. After that, it should be relatively easy to design a pricing policy to amortize the

investment in the stockpile with interest. This policy
could also be designed to encourage the further
extraction of helium from helium-lean natural gas
streams beyond 1985.

III. POLICY ALTERNATIVES AND IMPLEMENTING OPTIONS

Evaluation Criteria

Preceding sections of this report have described
the helium situation and enumerated the problems
associated with it. This section presents a range of
policy alternatives and suggests possible ways of
effecting the several policies. The Committee's charge
extends further, however, to analyzing the implications
of each policy alternative. The various types of
implications that are deemed worthy of analysis are
referred to as <u>evaluation criteria</u>.

Four standard evaluation criteria are generally
applied to the evaluation of public-sector undertakings:
(1) net economic benefits; (2) equity in the
contemporary and intergenerational distribution of
benefits and costs; (3) environmental protection; (4)
administrative feasibility. These will be defined and
illustrated below. In addition, this Committee has
defined two criteria that apply to particular issues of
the helium-conservation program: (5) protection of
future technological options; (6) appropriate division
of roles and responsibilities between the public and
private sectors to take greatest advantage of the
special capabilities of each.

While not all federal programs use a set of formal
criteria for evaluating and ranking policy or project
alternatives, the use of a formal, multiple-criterion
evaluation process is spreading. In the federal water-
resources area, the set of procedures referred to as
"the principles and standards" was approved by the
President in 1973 and is currently used by the Corps of
Engineers, the Bureau of Reclamation, and the Soil
Conservation Service. The principles and standards call
for criteria (1) and (3) above to be the primary
criteria for evaluating water projects, but (2) and (4)
are identified as factors that should be analyzed and
commented upon.

Net Economic Benefits

This is the criterion used in traditional benefit-
cost analysis, namely, the present value of the stream

of quantifiable benefits minus the present value of the stream of quantifiable costs associated with a program or project. If the program being evaluated produces an intermediate good, which is further processed and incorporated in other products, the annual net benefits are measured by the value added by program. This is the case with helium. However, it is difficult to project future prices for helium because future demands are uncertain. Today's market price of about $25 per Mcf will increase if demands are high enough to require production from higher-cost sources; however, it is not clear that the price will ever rise to the current cost of separation from air, that is, $2,500 to more than $3,000 per Mcf.

Equity in Contemporary and Intergenerational Distribution of Program Benefits and Costs

Equity is a matter of personal value judgment and cannot be scientifically decided. However, each program will have implications concerning the time pattern of possible benefits and costs, as well as who gets those benefits and who pays the costs at each point in time.

Environmental Protection

It is obvious that the maintenance of a high-quality physical environment is an important national policy objective. Any significant positive or negative effects of following a policy alternative, now or in the future, must be described at least in physical terms.

Administrative Feasibility

Policies differ in difficulty and cost of execution. The agencies and public involved must understand policy objectives and be motivated to cooperate in accomplishing them. The manpower should be sufficient to run the program and monitor its results effectively, and the program must be organized to avoid corruption. Other criteria being equal, administration should be as simple and consistent as possible.

Protection of Future Technological Options

Several helium-using technologies of great potential social importance are in the early development stage: superconducting devices and electric power transmission, magnetohydrodynamics, fusion power, and

others. Whether or not these technologies can be
brought to commercial fruition is not known. If they
are, it is not clear how important the cost of helium
will be in the determination of this possibility.
Storing helium today is a way of reducing the risk that
one or more of these technologies might be ruled out by
high future helium costs.

Appropriate Public-Private Roles

The federal government has played a vital role in
the history of helium production. The private sector
has played an important but smaller role to date, but
the Committee takes the view that the private sector
could play a greater role and could do so efficiently if
some of the government-induced uncertainty were removed.
Each policy alternative can be partially assessed in
terms of the extent to which it taps the special
capabilities of each sector.

Possible Components of a Helium Conservation Program

This subsection points out certain concrete steps
for the greater conservation of helium that might be
incorporated into a new helium program. These steps
deal only with the physical aspects of helium separation
and storage. Policy components relating to market
incentives, financial arrangements, legal
considerations, debt management, and taxes are reviewed
in the following Section.

Step I: Stop current venting of helium; to be
accomplished by federal government storage of this
helium, either after purchase by the federal government
or with ownership remaining private. This step would
stop the current economically and physically inefficient
wastage of helium that is being separated from natural
gas in the production of other products. The real
economic marginal cost of storing this gas would be
almost zero.

Step II: Designate helium currently stored in the
Cliffside field as a "national strategic reserve" to be
held against the possibility of major technologial
changes that would vastly expand the demand for helium
or that might require quantities in excess of current
production in private inventories. The "strategic
reserve" allows the federal government to bear the costs
of guarding against critical future shortages, but under
conditions that will not threaten the orderly
development of private provision of helium. The rules

for possible future releases from this strategic reserve would need to be set out very clearly and occasions for release from the reserve restricted to situations of major national significance.

Step III. Reactivate idle helium-separation plants. The next-lowest cost source of helium after the cessation of venting would be to reactivate these plants, which are located on gas streams from fields still under production. Because the capital costs of these plants are sunk and have been written off financially, the incremental cost of these supplies would still be very low, probably in the neighborhood of $7 per Mcf.

Step IV: Build new helium separation plants on helium-rich gas streams. Should current or future events make additional rates of separation and storage desirable, new plants would be required. These plants would be considerably more costly than those mentioned in Step III. Consideration should be given to the question of whether or not these plants should be multiple-product or helium-only producing plants.

Step V: Reserve helium-rich natural-gas fields. This step could refer to undeveloped natural-gas fields already in the public domain--fields producing under leases in the public domain, private undeveloped fields, and private producing fields.

The real costs involved in reserving fields that otherwise would be brought into production consist of increased production and transmission costs involved in the substitution of alternative supplies and any net benefit lost if some demands for natural gas are not met as a result of reductions in collection.

Insofar as reservation requires the cessation of production from currently active fields, legal liabilities for compensatory payments to private producers would exist.

Under present national natural-gas supply conditions, it appears that some of the short-term costs of such a program could be rather high. The longer-term costs would be noticeably lower as additional helium-lean sources are developed.

Policy Inducements

Strategies, Criteria, and Inducements

The actions outlined above, referred to as components of a helium-conservation program, imply objectives related to the quantity of helium that it might appear desirable to store for future use. They are strategies that might be adopted to produce different levels of conservation of the helium that exists in helium-rich natural-gas fields of the nation.

Now we turn to the policy inducements that might be employed to implement these strategies and we attempt to indicate how each might or might not accord with the evaluation criteria laid out above.

Tax and Other Incentives to the Private Sector

Tax incentives and disincentives could be established to encourage conservation by the private sector. Tax incentives might include (a) allowing companies to charge off as current expenses the costs of helium separation and storage; (b) working out an arrangement with local governments whereby companies are not taxed on helium they have put in storage; (c) a tax on helium wasted or dissipated into the atmosphere from helium-rich natural-gas streams, and (d) an investment tax credit on helium-separation facilities. These incentives would encourage the private sector to conserve helium, would not require direct government intervention in the helium market, and would not involve large direct costs to the government. Administration of these strategies would be somewhat more complex than that involved in direct intervention. How well the criteria of national economic welfare (net economic benefits), intergenerational equity, and protection of future technological options could be met with tax incentives alone would seem to depend on how much helium would be stored by private industry. It should be noted that there would be no end-use control, either in time or by nature of the use (dissipative, nondissipative; strategic, nonstrategic).

As environmental protection appears to be a useful criterion in evaluating helium-conservation options only as a consequence of high-energy use associated with atmospheric extraction, it may be dismissed for now.

Other incentives to private conservation could include government loans at subsidized interest rates, or government guarantees on private loans, for the

37

construction of helium-separation facilities, the provision of free storage in government storage facilities, positive government intervention in royalty litigation so that private conservation would not be blocked by the demands of owners of natural-gas mineral rights, and a government guarantee that it will not intervene directly in the helium market by selling from its stockpile or strategic reserve, either to private parties or government agencies, except under clearly specified and exceptional conditions.

The simplest, most immediate positive step that government could take to encourage private conservation would be to accept for storage helium now being vented, under an escrow type of agreement designed to prejudice the case of neither side in the current litigation.

Inducements to Increase Conservation by Government

To obtain the maximum economic benefits to the nation and an optimal level of transfer of wealth and preservation of technological opportunity to future generations, it may be necessary to increase the direct role of government in helium storage and conservation. There are several means by which this could be done. The government could resume, and expand, its helium-purchase program. It could build additional helium-separation plants of its own. The burden of the discount rate as a criterion of performance could be eliminated and the present "debt" to the U.S. Treasury written off. Remaining helium-rich natural-gas fields could be purchased by the government. These options are available within the existing frame of ownership of helium and its host (natural gas). Incidentally, helium could be purchased from the air-products industry on long-term contract, thus stimulating that industry to maximum separation of helium as a co-product.

An expanded role for government might involve changes in the institutional framework that now governs helium disposition. For example, because the federal government already has title to helium in natural gas on federal lands, leases for natural-gas exploitation could require that the gas streams be made available for helium separation, by either private or government plants, and lease sales could be deferred until such plants were in operation. Further, the government could reserve all "nondepleting" helium-rich fields until provisions are made for helium extraction. It could establish controls even on privately owned gas fields, curtailing or deferring their depletion until the contained helium can be removed and stored. It could

38

require that helium, above a specified low-level concentration, be separated from natural-gas streams and delivered to storage at producer expense; this would raise the price of natural gas to its consumers, who then would be paying an indirect tax to finance extraction and storage of helium.

The government can effect conservation on the demand side as well as on the supply side, by establishing end-use controls, designed (a) to minimize loss of helium by discouraging dissipative use, and (b) to limit the nonessential use of helium. To implement end-use controls and increase administrative feasibility, the government might assert a right to be the sole buyer and seller of helium.

Choosing a Policy Alternative

As we have suggested earlier in this report, the choice of a conservation policy alternative depends on a number of factors, foremost among which is an assessment of the ratio between the anticipated future benefits, about which there is great uncertainty, and the calculable present costs of the program under consideration. The Committee is convinced that the potential benefits are large compared to the present costs of conserving helium.

There is a consensus that the current practice of venting crude helium is unwise and inappropriate and should be halted. Quite probably, the current Cliffside stockpile should be turned into a national strategic reserve, the level of which should be the subject of further definitive study. Incentives to encourage helium conservation by private industry seem appropriate, though it is doubtful that any combination of incentives will produce the level of conservation desired. The government may need to take an aggressive role to resolve the legal and market uncertainties currently discouraging helium conservation. There are different opinions as to how such a problem should be analyzed, whether quantitatively by standard economic theory or qualitatively by nonmarket valuations, and there is some uncertainty as to the economic and social utility of the options discussed. Further definitive studies to resolve these uncertainties are clearly indicated.

The Committee has gathered and attempted to analyze a great deal of information. It has tried to present that information in useful form and has suggested alternative objectives and strategies for achieving different levels of helium storage and conservation.

The Committee recognizes certain institutional and political barriers to the implementation of the alternatives presented. The hardening of the government position relative to the Helium Act of 1960 because of the existing litigation appears to make it impossible to use that Act as a legal vehicle for a desirable national helium conservation program. Consequently, new legislation will be needed. Any new legislative effort will face the problem of acceptance, by the consumers of natural gas from which helium is extracted, by all taxpayers of the costs of benefits that will accrue mainly to those of the future, and by private industry whose appropriate role will require definition.

New sources of low-cost helium are conceivable but not likely. Consequently, there is a strong case for building a substantial government-owned strategic reserve of helium for use in the next century. There is also a good case for encouraging private industry to separate and store helium. How large the reserve should be and in what form, is a matter for policy makers to decide. A more detailed study is needed as a basis for these choices. In the meantime, continued atmospheric venting of helium from natural-gas separation plants is contrary to the national interest. The results of this preliminary analysis lead clearly and unambiguously to the conclusion that the venting of separated helium to the atmosphere, either directly or indirectly, should be stopped forthwith. Moreover, because of the massive uncertainties both in the long-term helium supply and in the national need, all appropriate measures should be taken in the near term to: seek ways to conserve the helium present in helium-rich natural gas accessible to the helium pipeline and the Cliffside storage field; hold government-owned helium stored in Cliffside as a strategic reserve; and induce or provide incentives for the private sector to store helium while it is readily available.

REFERENCES

Committee on Resources and Man. 1969. Resources and man. Preston Cloud [Chmn.]. National Research Council, National Academy of Sciences. W. H. Freeman and Company, San Francisco. pp. 10-11.

Energy Research and Development Administration. 1975. The energy related applications of helium and recommendations concerning the management of the federal helium program. A Report to the President and Congress of the United States. ERDA-13, April 11, 1975. 112 pp. plus 8 appendixes.

Exxon. 1976. U.S. oil and gas potential. Unpublished report by Exploration Department, Exxon Co. USA, March 1976, 10 pp. In Comparison of estimates of ultimately recoverable quantities of natural gas in the U.S. Gas Resource Study No. 1, April 1977. Potential Gas Agency, Colorado School of Mines, Golden, Colorado.

Laverick, Charles. 1975. Helium - Its storage and use in future years: Summary report. ANL/EE-75-2. Argonne National Laboratory, Argonne, Illinois. 15 pp.

Miller, Betty M., H. L. Thomsen, G. L. Dolton, A. B. Coury, T. A. Hendricks, F. E. Lennartz, R. B. Powers, E. G. Sable, and K. L. Varnes. 1975. Geological estimates of undiscovered recoverable oil and gas resources in the United States. Geological Survey Circular 725. Prepared for the Federal Energy Administration. USGS, Reston, Virginia.

Miller, Richard D., and Joe K. Tipton. 1977. Identified helium resources of the United States, 1977. U.S. Department of the Interior, Bureau of Mines, Washington, DC. 37 pp.

Moore, B. J. 1976. Helium resources of the United States, 1973. U.S. Bureau of Mines Information Circular 8708. U.S. Department of the Interior Library, Washington, DC. 18 pp.

Morgan, John D. 1977. Statement for Hearings on H. Res. 91, which expresses the sense of the House of Representatives that the President should direct the Secretary of the Interior to conserve helium, which is now being extracted from natural gas and then wasted into the atmosphere. Before the Subcommittee on Mines and Mining of the House Committee on Interior and Insular Affairs. Acting Director, Bureau of Mines, U.S. Department of the Interior, Washington, D.C., September 20. 6 pp. plus 2 appendixes.

Physics Survey Committee. 1972. Physics in perspective, Volume I. D. A. Bromley [Chmn.] National Research Council, National Academy of Sciences, Washington, D.C. pp. 42-43.

BIBLIOGRAPHY

Akerman, R.A., R.I. Rhodenizer, and C.O. Ward. 1977. A
    superconducting field winding subsystem for a 3,000 h.p.
    homopolar motor. IEEE Trans. on Magnetics, MAG-13(1)
    January.

Appleton, A.D. 1974. Superconducting DC machines: Concerning
    mainly civil marine propulsion but with mention of
    industrial applications. Invited Review Paper J8.
    Proceedings of the 1974 Applied Superconductivity
    Conference. IEEE Trans. on Magnetics, MAG-11(2):633, et
    seq. (March 1975).

Appleton, A.D., T.C. Bartram, R. Potts, and R.W. Watts.
    1977. Superconducting D.C. machines - A1MW propulsion
    system - Studies for commercial ship propulsion. IEEE
    Trans. on Magnetics, MAG-13(1) January.

Atherton, D.L., and A.R. Eastham. 1974. Superconducting
    Maglev and LSM development in Canada. Invited Review
    Paper J7. Proceedings of the 1974 Applied
    Superconductivity Conference.

Batzer, T.H. et al. 1974. Conceptual design of a mirror
    reactor for a fusion engineering research facility.
    August 28, 1974 UCRL-51617.

Baumol, W.J. 1970. On the discount rate for public projects.
    In Public expenditure and policy analysis. R.H. Haveman
    and J. Margolis [Eds.]. Markham Publishing Company,
    Chicago.

Bogner, G. Programs on large scale applications of
    superconductivity in the Federal Republic of Germany.
    In Superconducting machines and devices: Large systems
    applications.   Simon Foner and Brian B. Schwartz [Eds.].
    Advanced Study Institute Series B. Physics Plenum Press.

Boom, R.W. et al. 1974. Superconductive energy storage for large systems. Proceedings of the 1974 Applied Superconductivity Conference.

Cardwell, Louis Ernest. 1977. Helium resources on federal lands, 1977. U.S. Department of the Interior, Bureau of Mines, Washington, D.C. 16 pp.

Clark, Sherman H. and Frank E. Walker. 1971. The long range demand for helium. Stanford Research Institute Project ECC-8417. Prepared for Cities Service Helex Inc., Northern Helex Co., Phillips Petroleum Co. (Client Private). Menlo Park, California. 318 pp.

Committee on Resources and Man. 1969. Resources and Man. Preston Cloud [Chmn.]. National Research Council, National Academy of Sciences. W.H. Freeman and Company, San Francisco. pp. 10-11.

Cooper, R. 1974. Presentation at the USAED-DCTR Power Supply and Energy Storage Review Meeting, March 5-7, 1974. USAF Program. USAEC, Washington-1310. pp. 207-245.

Deaton, W.M. and R.D. Haynes. 1961. Helium production at the Bureau of Mines Keyes (Oklahoma) plant. Bureau of Mines Information Circular 8018. U.S. Department of the Interior, Bureau of Mines, Washington, D.C. 16 pp.

Derrick, M. et al. The Argonne National Laboratory--Carnegie Institute of Technology 25 cm bubble chamber superconducting magnet system. Proceedings International Conference on Instrumentation for High Energy Physics, Stanford, California, September 9-10, 1966.

Durant, W. and A. 1968. The lessons of history. Simon and Schuster. New York. 116 p.

Evans, David. 1970. Helium supply of the United States. The Helium Symposium Proceedings, March 23-24, 1970. Washington, D.C.: The Helium Society.

Exxon. 1976. U.S. oil and gas potential. Unpublished report by Exploration Department, Exxon Co. USA, March 1976, 10 pp. In Comparison of estimates of ultimately recoverable quantities of natural gas in the U.S. Gas Resource

Study No. 1, April 1977. Potential Gas Agency, Colorado School of Mines, Golden, Colorado.

Federal Helium Conservation Program. 1969. Hearings before the subcommittee on mines and mining of the Committee on Interior and Insular Affairs, September 1969.

Ford Foundation. 1974. Exploring energy choices. A preliminary report of the Ford Foundation's Energy Policy Project. S. David Freeman [Dir.]. Ballinger, Cambridge, Mass. 141 pp.

The future revised. 1976. Wall Street Journal (April 15, 1976).

Hirshleifer, J. and D.L. Shapiro. 1970. The treatment of risk and uncertainty. In Public expenditure and policy analysis. R.H. Haveman and J. Margolis [Eds.]. Markham Publishing Company, Chicago.

Howland, H.R. and J.K. Hulm. 1974. The economics of helium conservation. Westinghouse Research Laboratories Report 74-8C53-HELIUM-R1, December. (Argonne National Laboratory Contract #31-109-38-2820), also in Hearings before the House Subcommittee on Energy Research, Development, and Administration, May 7, 1975.

Ibbotson, Roger B. and Rex A. Sinquefield. 1977. Stocks, bonds, bills and inflation: The past (1926-1976) and the future (1977-2000). Financial Analysts Research Foundation.

Kamper, R.A. 1974. Review of superconducting electronics. Invited Plenary Session Paper B6. Proceedings of the 1974 Applied Superconductivity Conference.

Kantrowitz, Arthur and Jacob Zar. 1970. MHD electric power and its need for helium. Helium Symposium of the Helium Society, 173 K Street, N.W., Washington, D.C.

Krupka, M. C. and E. F. Hammel. 1977. Energy, helium and the future. Submitted to the Symposium on Alternative Energy Sources, Miami Beach, Florida, December 5-7, 1977. The Los Alamos Scientific Laboratory identifies this article as work performed under the auspices of USERDA. 17 pp.

Laverick, Charles. 1968. High field superconductor technology. IEEE Spectrum, Vol. 4, No. 4, April 1968.

Laverick, Charles. 1971. Helium conservation and its relevance to future technology. 1971 Helium Society Symposium.

Laverick, Charles [Ed.]. 1971. The U.S. helium program and controversy. Cryogenics, December 1971.

Laverick, Charles. 1973. Helium study notes: Part I, 260 pp.; Part II, 226 pp.; Part III, 103 pp.; Part IV, 336 pp.; Part V, 30 pp.; Part VI, 36 pp. Argonne National Laboratory, Argonne, Illinois.

Laverick, Charles. 1974. Helium - its storage and use in future years: Preliminary report. ANL/EE-75-1. Argonne National Laboratory, Argonne, Illinois. With an appendix by David M. Evans on helium supply and an appendix by H. R. Howland and J. K. Hulm on the economics of helium conservation. November 1974.

Laverick, Charles. 1975. Helium - its storage and use in future years: Summary report. ANL/EE-75-2. Argonne National Laboratory, Argonne, Illinois. 15 pp.

Laverick, Charles. 1976. Part I: Present status and future prospects for superconductivity in the USA. Part II: The present status of applied superconductivity in the USA. Part III: U.S. energy policy. Presented at the USSR Conference on the Technical Applications of Superconductivity in Alushta, USSR, September 15 to October 4, 1975. 74 pp.

Laverick, Charles. 1976. Applied superconductivity in the USSR. Conference on the Technical Applications of Superconductivity in Alushta, USSR, September 15 to October 4, 1975. 63 pages.

Laverick, C. and G. Lobell. 1965. A large, high-field superconducting magnet system. Rev. Sci. Inst. 36, 825.

Laverick, C. and J. Powell. 1973. Applied superconductivity in the CTR program. A report to the USAEC Division of Controlled Thermonuclear Research. August 28, 1973.

Lipper, Harold W. 1970. "Helium." In Mineral facts and problems. U.S. Bureau of Mines reprint from Bulletin 650. Washington, D.C.: U.S. Government Printing Office.

Midwest Research Institute. 1977. Comprehensive investigation and report on helium uses. James D. Maloney [Dir.]. Final report on MRI Project No. 4157-D. U.S. Bureau of Mines, Washington, D.C. 142 pp.

Miller, Betty M., H. L. Thomsen, G. L. Dolton, A. B. Coury, T. A. Hendricks, F. E. Lennartz, R. B. Powers, E. G. Sable, and K. L. Varnes. 1975. Geological estimates of undiscovered recoverable oil and gas resources in the United States. Geological Survey Circular 725. Prepared for the Federal Energy Administration. U.S. Geological Survey, Reston, Virginia.

Miller, Richard D. and Billy J. Moore. 1977. Estimated helium in identified and undiscovered supply of natural gas in the United States from 1977 through 2020. U.S. Department of the Interior, Bureau of Mines, Washington, D.C. 37 pp.

Miller, Richard D. and Joe K. Tipton. 1977. Identified helium resources of the United States. U.S. Department of the Interior, Bureau of Mines, Washington, D.C. 37 pp.

Moir, R.W. and C.E. Taylor. 1972. Magnets for open-ended fusion reactors. For the Symposium on the Technology of Controlled Thermonuclear Fusion Experiments and the Engineering Aspects of Fusion Reactors, Austin, Texas, November 20-22, 1972. Also UCRL-74326.

Montgomery, D. Bruce. 1969. Solenoid magnet design. J. Wiley and Son, New York.

Moore, B.J. 1976. Helium resources of the United States, 1973. U.S. Bureau of Mines Information Circular 8708. U.S. Department of the Interior, Washington, D.C. 18 pp.

Morgan, John D. 1977. Statement for Hearings on H. Res. 91, before the subcommittee on mines and mining of the House Committee on Interior and Insular Affairs, September 20, 1977. U.S. Department of the Interior, Bureau of Mines, Washington, D.C. 35 pp.

Morton, Rogers C.B. 1973. Termination of helium purchase contracts. U.S. Department of the Interior, Washington, D.C. 17 pp.

Murray, B. and M.E. Davies. 1976. Detente in space. Science 192(4244) (June 11, 1976).

Ohtsuka, T. and Y. Kyotani. 1974. Superconducting levitated high speed ground transportation project in Japan. Invited Paper J2. Proceedings of the 1974 Applied Superconductivity Conference.

Parker, J.H., Jr., et al. 1975. A high speed superconducting rotor. Westinghouse Research Paper J9. Proceedings of the 1974 Applied Superconductivity Conference. In IEEE Trans. on Magnetics, MAG-11(2):640 (March 1975).

Physics Survey Committee. 1972. Physics in perspective, Vol. I. D. A. Bromley [Chmn.]. National Research Council, National Academy of Sciences, Washington, D.C. pp. 42-43.

Pierce, A.P. et al. 1964. Uranium and helium in the panhandle gas field, Texas and adjacent areas. Geological Survey Professional Paper 454-G U.S. From shorter contributions to General Geology. Prepared on behalf of the U.S. Atomic Energy Commission. Washington, D.C.: U.S. Government Printing Office.

Potential Gas Agency. 1977. Comparison of estimates of ultimately recoverable quantities of natural gas in the U.S. Gas Resource Study No. 1. April 1977. Potential Gas Agency, Colorado School of Mines, Golden, Colorado.

Powell, J. Magnetic levitation - High speed ground
     transportation. In Superconducting machines and devices
     - Large systems applications. Simon Foner and Brian B.
     Schwartz [Eds.]. NATO Advanced Study Institute Series B.
     Physics Plenum Press.

Preston, Lee. 1969. The federal helium program. Testimony to
     the subcommittee on economy in government, Joint
     Economic Committee. September 23, 1969.

Preston, Lee and David Brooks. 1974. The federal helium
     program. Report to Council of Economic Advisers on
     December 1, 1974.

Price, Charlotte Alber. 1971. The helium conservation
     program of the Department of the Interior. Environmental
     Affairs (publication of the Environmental Law Center,
     Boston College Law School, Brighton, Massachusetts
     Institute of Technology, Cambridge, Massachusetts. 4 pp.

Price, P. J. and R. L. Garwin. 1977. APS (POPA) statement
     on helium conservation. Final statement of American
     Physical Society. Massachusetts Institute of
     Technology, Cambridge, Massachusetts. 4 pp.

Proceedings of the 1974 Applied Superconductivity
     Conference, Oakbrook, Illinois. In IEEE Trans. on
     Magnetics. Mag-11(2) (March 1975). 891 pp.

Proceedings of the 1976 Applied Superconductivity
     Conference. In IEEE Trans. on Magnetics. MAG-13(1)
     (January 1977). 940 pp.

Rabinowitz, Mario. 1975. The Electric Power Research
     Institute's role in applying superconductivity to future
     utility systems. Invited Plenary Session paper A4.
     Proceedings of the 1974 Applied Superconductivity
     Conference.

Ray, Dixy Lee. 1973. The nation's energy future: A report to
     Richard M. Nixon. Submitted December 1, 1973. WASH 1281.
     U.S. Atomic Energy Commission, Washington, D.C.

Reitz, J.R. and R.H. Borcherts. 1974. U.S. Department of
     Transportation program in magnetic suspension (repulsion

concept). Paper J3. Proceedings of the 1974 Applied
Superconductivity Conference.

Ross, Robert S. and Philip F. Myers. 1970. A transportation
system for the 1970s and beyond. The Helium Society
Symposium Proceedings. Goodyear Aerospace Corporation,
Akron, Ohio. March 1970.

Seibel, C.W. 1968. Helium, child of the sun. University Press of
Kansas, Lawrence, Kansas.

Smith, J.L. et al. 1974. Superconducting rotating machines.
Invited Review Paper B4. Proceedings of the 1974 Applied
Superconductivity Conference.

Stockholm International Peace Research Institute. 1975.
World armaments and disarmament: SIPRI Yearbook 1975.
MIT Press, Cambridge, Mass.

Stores, John. 1976. The strategic nuclear arms race. Impact
of science on society. 28(1/2).

Superconducting machines and devices: Large systems
applications. 1974. S. Foner and B. Schwartz [Eds.].
Proceedings of the NATO Advanced Study Institute,
Entreves, Italy. Series B Physics. Plenum Press, New
York. 692 pp.

Tang, C.H. et al. 1974. A review of the Magneplane Project.
Paper J6. Proceedings of the 1974 Applied
Superconductivity Conference.

Tuck, J.L. 1971. On the outlook for controlled fusion power
with notes on its relation to helium consumption. The
Helium Symposium Proceedings, May 1971.

U.S. Atomic Energy Commission, Division of Controlled
Thermonuclear Research. 1973. Fusion power, an
assessment of ultimate potential. February 1973.

U.S. Comptroller General. 1963. Examination of procurement
of crude helium for the helium conservation program
under negotiated fixed-price contracts, U.S. Department

of the Interior, Bureau of Mines. Report to the Congress of the United States, January 1963.

U.S. Comptroller General. 1965. Negotiation of non-competitive fixed price contracts for procurement of helium bearing gas without adequate determination of reasonableness of prices, U.S. Department of the Interior, Bureau of Mines. Report to the Congress of the United States, June 1965.

U.S. Comptroller General. 1969. Review of the government's program to supply current and future helium requirements, U.S. Department of the Interior, Bureau of Mines, B-114812. Report to the Committee on Interior and Insular Affairs, House of Representatives, September 10, 1969.

U.S. Congress. 1960. An Act To Amend the Helium Act of March 3, 1925, as amended, Public Law 86-777, 86th Congress, H.R. 10548, September 13, 1960. U.S. Senate and House of Representatives, Washington, D.C. 6 pp.

U.S. Court of Claims. 1975. Appendix to petition for a writ of certiorari to the United States Supreme Court, Northern Helex Company, Petitioner, v. The United States of America, Respondent. Byron S. Adams Printing, Inc., Washington, D.C. 318 pp.

U.S. Court of Claims. 1975. Defendant's exceptions to the trial judge's findings of fact. Northern Helex Company v. The United States. No. 454-70 (filed March 24, 1975). Edward J. Friedlander, Attorney, Civil Division, Department of Justice. Irving Jaffe, Acting Assistant Attorney General, Civil Division, Department of Justice. John D. Trezise, Attorney, Department of the Interior.

U.S. Court of Claims, Trial Division. 1975. Northern Helex Company v. The United States. No. 454-70 (filed December 3, 1974). Report of the Trial Judge to the Court. In the Supreme Court of the United States, October Term, 1975. Byron S. Adams Printing, Inc., Washington, D.C. pp. 42-298.

U.S. Court of Claims. 1975. Opinion of the U.S. Court of Claims. Northern Helex Company v. The United States. No.

454-70 (decided October 22, 1975). Before Chief Judge Durfee.

U.S. Department of the Interior. 1976. Briefing on helium resources, supply and demand, and conservation. Presented to the Director, Bureau of Mines. USBM, Division of Helium, Washington, D.C. 46 pp.

U.S. Department of the Interior, Bureau of Mines. 1947. Helium: Bibliography of technical and scientific literature from its discovery (1868) to January 1, 1947 by Henry P. Wheeler, Jr., and Louise B. Swenarton. Bulletin 484.

U.S. Department of the Interior, Bureau of Mines. 1969. Helium: Bibliography of technical and scientific literature, January 1, 1947, to January 1, 1962. A supplement to Bulletin 484.

U.S. Department of the Interior, Bureau of Mines. 1969. Helium: Bibliography of technical and scientific literature, 1962. Includes papers on alpha-particles. Circular 8398.

U.S. Department of the Interior, Bureau of Mines. 1970. Helium: Bibliography of technical and scientific literature, 1963. Includes papers on alpha-particles. Circular 8467.

U.S. Department of the Interior, Bureau of Mines. 1970. Helium: Bibliography of technical and scientific literature, 1964. Includes papers on alpha-particles. Circular 8489.

U.S. Department of the Interior, Bureau of Mines. 1971. Helium: Bibliography of technical and scientific literature, 1965. Includes papers on alpha-particles. Circular 8523.

U.S. Department of the Interior, Bureau of Mines. 1972. Environmental statement, termination of helium purchase contracts, DES 72-6. Draft May 16, 1972, final November 1972.

U.S. Department of the Interior, Bureau of Mines. 1976.
Helium. In Committee print, Ninety-Fourth Congress,
2nd Session. Prepared by Congressional Research Service,
Library of Congress, September 1976.
U.S. Government Printing Office, Washington, D.C.

U.S. Department of the Interior, Bureau of Mines, Division
of Helium. 1977. Briefing charts on federal helium
program, presented to Interagency Helium Committee.
Washington, D.C. 51 pp.

U.S. Department of the Interior, Bureau of Mines, Division
of Helium. 1977. Bureau of Mines helium facilities.
Washington, D.C. 16 pp.

U.S. Department of the Interior, Office of Economic
Analysis. 1972. Analysis of the helium program relative
to the proposed termination of helium purchase contracts
as set forth in the under-secretary's termination
decision of January 26, 1971 and the Bureau of Mines'
draft environmental impact statement of May 1972.

U.S. Department of the Interior, Secretary of the Interior.
1976. Report to the Congress on matters contained in the
Helium Act (Public Law 86-777), Fiscal Year 1976.
October 1976.

U.S. District Court, Kansas District. 1973. National Helium
Corporation et al. v. Rogers C.B. Morton, Secretary of
the Interior and Elburt F. Osborn, Director, Bureau of
Mines. Order by Judge Frank G. Theis. June 11, 1973.

U.S. Energy Research and Development Administration. 1975A.
The energy related applications of helium (ERDA-13).
E.F. Hammel [Dir.]. A report to the President and the
Congress of the United States. Office of the Assistant
Administrator for Conservation, U.S. Energy Research and
Development Administration, Washington, D.C. 112 pp.
plus appendices.

U.S. Energy Research and Development Administration. 1975B.
A national plan for energy research, development and
demonstration: Creating energy choices for the future.
June, ERDA 48, Vol. 1.

U.S. Federal Energy Administration. 1974. Project
    Independence Report, p. 430.

U.S. House of Representatives. 1964. Hearings before a
    subcommittee of the Committee on Appropriations, U.S.
    Senate. Eighty-Eighth Congress, First Session, June 30,
    1964. H.R. 5279.

U.S. House of Representatives. 1960. (Helium) Hearings
    before the subcommittee on mines and mining of the
    Committee on Interior and Insular Affairs. Eighty-Sixth
    Congress, Second Session, 1960. H.R. 8440.

U.S. House of Representatives. 1970. Federal helium
    conservation program. Hearings before the subcommittee
    on mines and mining of the Committee on Interior and
    Insular Affairs. Ninety-First Congress, First Session,
    September 15-16, 1969. Serial No. 91-11. U.S. Government
    Printing Office, Washington, D.C. 402 pp.

U.S. House of Representatives. 1975. The energy related
    applications of helium. Hearing before the subcommittee
    on energy research, development and demonstration of the
    Committee on Science and Technology. Ninety-Fourth
    Congress, First Session, May 7, 1975 (No. 10).
    Washington, D.C.: U.S. Government Printing Office. 576
    pp.

U.S. House of Representatives. 1977. Helium fund. Report No.
    95-392. Washington, D.C.: U.S. Government Printing
    Office. pp. 40-41.

U.S. Senate. 1976. Helium conservation. Senate Report 94-
    808. From Mr. Metcalf to accompany Committee on Interior
    and Insular Affairs. S.R. 253. Ninety-Fourth Congress.
    Second Session, May 11, 1976.

U.S. Senate. 1977. Statements on introduced bills and joint
    resolutions. Congressional Record. September 19, 1977:
    S15163-15165.

U.S. Senate. 1977. A Bill, Title I - Helium Act Amendments,
    S.2109. Ninety-Fifth Congress, First Session, September
    19, 1977. U.S. Government Printing Office, Washington, D.C. 9 pp.

U.S. Senate. 1977. Helium fund. Report No. 95-276, Calendar 256. Washington, D.C.: U.S. Government Printing Office. p. 24.

University of Wisconsin, Engineering Experiment Station, College of Engineering. Wisconsin superconductive energy project. Vol. 1, July 1, 1974. Final report to NSF.

University of Wisconsin, Revision Nuclear Engineering Department. Potential CTR requirements for helium up to the year 2020, April 1974.

Ward, Dwight E. and Arthur P. Pierce. Helium. In United States mineral resources. U.S. Geological Survey Professional Paper 820. Washington, D.C.: U.S. Government Printing Office. p. 285-290.

World Electrotechnical Congress. 1977. Scientific program, Moscow, June 21-25, 1977.

Zumwalt, E. 1976. On watch. Quadrangle Press. Chicago.

National Academy of Sciences
National Research Council
Commission on Natural Resources
Board on Mineral and Energy Resources

Present
a Public Forum on

HELIUM:
PRESENT AND FUTURE NEEDS

November 20-21, 1977

Washington, D.C.

HELIUM STUDY COMMITTEE

Robert M. Drake, Jr. (Chairman), Vice President-Technology, Studebaker-Worthington, Inc.

William D. Carey, Executive Officer, American Association for the Advancement of Science

Earl F. Cook, Dean of Geosciences, Professor of Geography and Geology, Texas A&M University

Charles W. Howe, Professor of Natural Resources Economics, University of Colorado; Visiting Professor of Agricultural and Applied Economics, University of Minnesota

John K. Hulm, Manager, Chemical Sciences Division, Westinghouse Research and Development Center

Lester B. Lave, Professor and Head, Department of Economics, Carnegie-Mellon University

Franklin A. Long, Henry Luce Professor of Science and Society, Professor of Chemistry, Cornell University

F. Clayton Nicholson, Management Consultant, Northern Natural Gas Company

Michael Tinkham, Professor of Physics, Gordon McKay Professor of Applied Physics, Harvard University

Albert E. Utton, Professor, School of Law, University of New Mexico

Irvin L. (Jack) White, Professor of Political Science and Project Director, EPA Western Energy Technology Assessment, University of Oklahoma

CONTENTS

3

DAY I

<u>Welcome and Introduction</u>

ROBERT R. WHITE

Director, Academy Forum

The Academy Forum tends to focus on public issues about which there are many public questions, issues that are controversial, and issues that involve science. This meeting is a rather special kind of Forum, and my capacity in it is as part of the staff of the agencies of the Academy that are interested in this particular Forum. These agencies are: The National Research Council, the Commission on Natural Resources, the Board on Mineral and Energy Resources, and the Helium Study Committee that has recently been convened.

We have many kinds of public meetings in the Academy: symposia, colloquia, seminars, many informal meetings, and occasionally we have meetings that are something like hearings. The special purpose of this meeting is to assist the Helium Study Committee, which has a very difficult task to accomplish in a very short time. The key audience of this Forum is the Committee, which must prepare by mid-December a report for the Bureau of Mines that will shed some light on the problems of helium. This is an extraordinarily short length of time.

This meeting has been organized to get as much input from highly expert, highly relevant viewpoints as possible. The purpose of such a Forum is not to settle things or to come up with answers, but to review issues, to review uncertainties, to review data, and to review options. The members of the Helium Study Committee have been selected with great care by the regular processes of the Academy. Of particular importance to any of these efforts is the brightness, the impartiality, and the leadership of the Chairman. Your Chairman is Robert M. Drake, Jr., a

4

mechanical engineer, a member of the National Academy of
Engineering, a former Dean of Engineering at the University
of Kentucky, and now a Vice President for Technology for
Studebaker-Worthington, Incorporated.

5

Purpose of the Forum

ROBERT M. DRAKE, JR.

Chairman
Helium Study Committee

Today we have been working as a committee. We have made
some progress on our assignment, and we have developed some
understanding of the complexities of helium. We now wish to
listen to our speakers and audience tonight and tomorrow,
and to improve that understanding.

First, I would like to read the request from the
Interagency Helium Study Group to the National Academy of
Sciences that precipitated the forming of this Committee.

"The House Committee on Appropriations in its Report No.
95-392, on H.R. 7636, Department of the Interior and Related
Agencies Appropriation Bill for 1978, ordered a Joint Helium
Study by the Bureau of Mines and the Energy Research and
Development Administration, now the Department of Energy, to
be completed no later than January 31, 1978. The Senate
Committee on Appropriations concurred. The language of
those two reports briefly is as follows:

House of Representatives
Report No. 95-392

The Committee recommends recission of the
permanent contract authority to become available in
fiscal year 1978 [this relates to helium extraction
and storage]. Current helium reserves in storage
and in nondepleting reservoirs are 60 billion cubic
feet, or 100 years supply at current rates of
usage. The Committee is concerned that large
volumes of helium are being vented to the
atmosphere in natural gas recovery operations each
year, because to recover helium from the
atmosphere, if needed at a later time, currently is
considerably more expensive than recovery from

6

natural gas streams. The need for large quantities of helium in the next century depends on uncertain demand projections based on technological advances that may occur well into the future in areas such as superconducting power transmission, magnetic energy storage, and fusion.

Because of the uncertainty in future demands on the one hand, and the ease with which helium can presently be recovered and stored on the other, the Committee understands the concern of some in approving the recission of these funds. Therefore, the Committee directs that the Bureau and the Energy Research and Development Administration jointly enter into a helium study to be completed no later than January 31, 1978, to be available in time for consideration for the fiscal year 1979 budget. Such a study should update the estimates of demand and uses discussed in ERDA's report on helium dated April 10, 1975, and consider costs and benefits of various courses of action based on various demand scenarios. Changes to existing laws that would facilitate the conservation of helium should also be discussed and evaluated. The report should describe various alternative options and make recommendations for any changes deemed necessary in the federal helium program.

Senate
Report No. 95-276

The Committee recommends the recission of $47,500,000 in permanent contract authority for helium purchases, in agreement with the House. This recommendation, consistent with Congressional action in prior years, stems from uncertainty over the status of helium contracts cancelled by the Department and now in litigation. The Committee fully concurs in the House directive that the Bureau and the Energy Research and Development Administration jointly conduct a study aimed at developing a sound Federal policy on helium conservation, storage, and future needs.

"Although the Congress called for a joint study by the Bureau of Mines and ERDA, the Department of Defense and the National Aeronautics and Space Administration were invited to participate since those two agencies account for about 80 percent of the helium used by the Federal Government. Accordingly, a committee composed of designated

representatives of Mines, ERDA, DOD and NASA was
established. The Committee met on August 24, 1977, and,
among other things, concluded that an assessment of the
helium situation should be obtained from an independent,
non-Government organization, namely, the National Academy of
Sciences.

"The assessment by NAS should encompass the
philosophical, including socioeconomic and legal aspects of
further helium conservation. One way to view the helium
problem is as discussed below:

"Although there is an inexhaustible supply of helium in
the atmosphere--5,000 cubic miles--the low concentration--5
parts per million--makes recovery from that source
prohibitive in terms of economics and energy consumption
unless new extraction technology is developed. The only
economic source of helium at this time is natural gas. All
natural gas contains helium in concentrations far greater
than the atmosphere, and so long as natural gas is available
in sufficient quantities there will be no need to resort to
the atmosphere for helium.

"Natural gas is, of course, a finite resource. When all
natural gas supplies are gone, so will be helium, except for
atmospheric helium. It is a basic fact, therefore, that if
helium continues to be used in various technologies, its
source eventually will be the atmosphere.

"Thus, the first question that needs consideration would
be: In view of other national priorities, what could, or
what should, be done to delay the time when the Nation's
helium requirements must be obtained from the atmosphere,
which, in any event, is inevitable?

"If a decision is made to stretch supplies of helium
available from natural gas further into the future, how far
does the Nation go? Should every atom be saved at the
expense of some other National goal or goals? Or should
one-half, or one-fourth, or some other fraction of the atoms
of helium in natural gas be saved?

"What are the alternatives for accomplishing the helium
conservation goal deemed to be necessary and/or economically
feasible, considering current factors such as the already
huge helium conservation debt, legal entanglements involving
the value of helium at the wellhead, the Helium Act, tax
laws and any other factors believed to be pertinent?"

That letter resulted in the formation of our Committee,
and therein lies the reason for undertaking the study.
[Since then, these suggestions and the Committee's report

have been modified slightly by the wording of the contract agreement that became effective November 21, 1977.] Now the Committee already has a relatively large amount of information available to it, for example, the Argonne National Laboratory reports by Laverick, The Energy Related Applications of Helium (ERDA-13) directed by Hammel, the Westinghouse Report by Howland and Hulm, Senate Bill 2109, House Resolution 91, The American Physical Society statement by Price and Garwin, and an American Chemical Society study for which a report is not yet available.

What the Committee will try to do is to stress issues, uncertainties, and options in areas of technical, economic, legal, and institutional factors.

An Overview:
PAST AND PRESENT USES OF HELIUM

Harold W. Lipper
Retired Chief
Division of Helium, Bureau of Mines

Helium has been known since 1895 when Ramsay found it to
occur in minerals, confirming the 1868 discovery of it in
the sun by Lockyer. Immediately upon the discovery of
helium, scientists began to do things with it and wanted to
know more about its properties. This was the golden age of
science and many discoveries were being made. By 1905--
before it was discovered in a Kansas natural-gas deposit--
quite a bit was known about helium. It occurs in gases from
hot springs, the atmosphere, and many minerals. It is
present in the sun, other stars, nebulae, and meteorites.
It is generated spontaneously by radioactive decay. It is
known to be the lightest of all gases except hydrogen; it is
a single atom with a molecular weight of about 4; it
diffuses rapidly, even through quartz; it is chemically
inert; it is only slightly soluble in liquids, odorless,
tasteless, nonflammable, and nonpoisonous in the ordinary
sense.

Thermometry

Helium's first use was in the measurement of low
temperatures. Kelvin had first suggested the absolute scale
of temperature measurement in 1848. At that time, scales
used most commonly were based on the freezing and boiling
points of water. Kelvin wished to avoid relating the
temperature scale to the properties of any specific
material; instead, he proposed a scale where zero would be
the temperature at which the volume of a perfect gas at
constant pressure would be reduced to zero.

Hydrogen was first liquefied and used to fill
thermometers for low temperature measurements. Then,
Professor Karol S. Olszewski of Poland, in 1895, suggested
the use of helium in such a thermometer when he was the
first to attempt to liquefy helium and failed. He had

10

140 cc of helium Ramsay had sent him. In 1913 the Fifth
General Conference on Weights and Measures began work to
develop conversion tables so that helium could be used to
fill thermometers for low temperature measurements. Helium,
of all gases, behaved most nearly like a perfect gas.

Lifting Gas

     The second use of helium was for inflating airships. In
1920 the helium that had been recovered from natural gas in
U.S. Navy plants during World War I was used to inflate the
Navy's C-7. That airship operated as a prototype for larger
ships--specifically the Shenandoah--and was used for
learning how to handle an airship without valving or venting
its buoyant gas. Later, sister ships the Los Angeles, the
Akron, and the Macon were inflated with helium. Each of them
had a capacity of 6.5 million cubic feet (MMcf); each was 785
feet long and 133 feet in diameter. Cruising range for
these airships was over 6,000 miles at a cruising speed of
over 80 miles an hour. Five scout planes were carried on a
hangar deck. The planes were launched from and retrieved
with a trapeze arrangement slung below the airship. All of
these ships except the Los Angeles were lost in crashes.
Goodyear's first ship, Pilgrim, was inflated with Navy
helium in 1925. Later, helium was leased to Goodyear
because government regulations prevented its sale. In 1935,
a stratosphere balloon flight set a new altitude record of
over 72,000 feet. During World War II, helium-filled blimps
were used for antisubmarine patrol along both U.S. coasts
and for convoy escort in the Atlantic. By 1961, the U.S.
Navy was out of the blimp business although there were plans
for a nuclear-powered blimp to serve as a radar platform.

     After several hydrogen-caused explosions, the Weather
Bureau began using helium in its weather balloons in 1939.
By the mid 1960s, helium use was phased out except for
shipboard use and some special applications. One such
application was a grid of constant volume balloons, at
heights over 100,000 feet, for gathering weather information
in a horizontal plane instead of the normal vertical
information-gathering flights. Last winter I was told that
a large tethered balloon aloft in the Florida Keys was a
high altitude platform for electronic gear in an Air Force
project. There continues to be discussion about a new breed
of airship for carrying freight both here and in Germany.

     In the Pacific Northwest, helium is used in logging
operations; in rugged terrain a cable is rigged between two
supports--one uphill and one downhill. A helium-filled
blimp attached to the line supports the weight of the logs,
and they are winched down the slope. Few roads are required
and damage to the soil surface is minimized.

11

Some of the small but important helium uses are worth mentioning here for want of a better place to put them. Helium was used in the 1930s and 1940s to dilute anesthetics and to prevent operating-room explosions. There were instances where the cyclopropane oxygen anesthetic mixture used in those days exploded in a patient's lungs. Neon signs, which appeared in the 1930s, give off a yellow glow when filled with helium. A patent was issued in 1934 for filling tubes of telescopes and range finders with helium, which has a much lower refractive index than air. I think the Navy used some of these systems for aiming naval guns. In 1945 there was some publicity about helium use in airplane tires to reduce weight. The idea never caught on.

Synthetic Breathing Mixtures

In the 1930s, Dr. Alvan Barach of Columbia University began using helium-oxygen mixtures for the treatment of severe cases of asthma. Patients got instant relief. He showed movies of his treatment to a Congressional committee with the result that the 1937 Helium Act and subsequent amendments have provided that helium for medical purposes shall always be sold at a cost that will not prohibit its use for such a purpose.

Helium was proposed as a substitute for nitrogen in the space cabin atmosphere of the planned Air Force manned orbiting laboratory. The idea was to reduce spacecraft weight for long missions. Russia announced similar plans. However, the Air Force project was cancelled early in the development stage.

Diving

The use of helium in diving is somewhat related to the uses just mentioned. A patent on a respirable mixture of oxygen and helium was applied for in 1919 and granted in 1923. Beginning in 1925 the Bureau of Mines, the U.S. Public Health Service, and the U.S. Navy began a project at Pittsburgh on the use of such a mixture in diving operations. A small pressure chamber was built, and rats and guinea pigs were the first subjects. Later, a man-sized chamber was built that stayed in use into the 1960s. The first idea was to put a diver suffering from the bends into the pressure chamber to decompress while breathing a helium-oxygen mixture. Saturation diving was anticipated, but helium costing $35 to $50 a thousand cubic feet (Mcf) was considered too expensive. After completion of the initial work, the Navy Diving School was moved to Washington, D.C., where simulated dives were made to 500 feet.

The first use of the new technique came in 1939 when the USS Squalus sank in 240 feet of water. The Navy used conventional diving methods to rescue the 33 survivors. After that was done, helium-oxygen mixtures were used by Navy divers to raise the vessel.

In 1941, two dives using a helium-oxygen mixture were made to 440 feet in order to reach the sunken USS Conger. At the same time, work was going on in the British Navy involving dives to 540 feet in 1948 and by 1956 to 600 feet. The U.S. Navy limited its operational diving depths to 380 feet because of a lack of knowledge in computing and predicting decompression requirements at greater depths.

Hannes Keller, a Swiss, went to 700 feet for 10 minutes in 1962, and later made a dive to 1,000 feet in the open sea. That was the year that Captain George Bond, U.S. Navy, successfully demonstrated saturation diving in the open sea; and a California diving firm diver went below 400 feet to demonstrate to the offshore oil industry that diving below 200 feet was feasible. Scott Carpenter, the astronaut, made valuable contributions to diving in the Navy Sealab projects.

Progress in this area has been rapid. There are many firms engaged around the world in diving operations in connection with offshore oil and gas operations, and the Navy has a major installation in support of diving at Pensacola, Florida. Test dives have been made to a simulated depth of 1,700 feet for a period of six days. During that test, neon was substituted for helium to obtain conditions equivalent to breathing helium at a depth of 5,000 feet.

Helium as a Tracer

Analysis for small amounts of helium in other gases is a very simple procedure so that it was perhaps natural to think about injecting helium into oil and gas fields to determine the underground movement of injected gas. Work along these lines was done by the Bureau of Mines at locations in California and West Virginia in the late 1940s.

Leak Detection

As better means for helium analysis were developed, it was possible to detect smaller and smaller quantities with precision.

Consequently, helium is used for the detection of minute leaks or more often for the complete absence of any leaks. Methods used are so sensitive that leaks of less than 1 liter of helium over a period of 3,000 years can be detected. Helium leak detection is used for quality control of welds and joints on vacuum systems, refrigerator spacecraft piping, nuclear reactor components, and other fabricated items where freedom from leaks is essential.

Chromatography

Helium is used as a carrier gas in a column which may be filled with different materials and operated over various temperature ranges. Substances being analyzed are separated in the column and are progressively displaced by the helium carrier gas. The method is rapid, precise, and relatively simple, but it is fairly costly. There are probably more than 50,000 such units in operation and--at least in the late 1960s--more than 90 percent used helium. Application is rather broad for quality control in food, drug, petrochemical, paint, and other industries as well as in various other analytical laboratories.

Welding and Metallurgy

The first patent for welding in an atmosphere of helium was granted to Henry Hobart in 1930. An early application of this use was made during World War II by Northrup Aircraft to repair and salvage faulty magnesium castings. In contact with oxygen, magnesium will burn with intense heat. Argon had seemed the ideal choice because, like helium, it is inert; however, the argon available at that time was wet because it was pumped with water-lubricated compressors, whereas--by virtue of the extraction process-- helium was essentially dry. Imagine our surprise when we learned that although the helium was dry, as it came from the purification equipment, the tank cars and cylinders we put it into were wet from previous hydrostatic testing. We managed to dry the containers and, for a number of years, one of my duties was to check shipments to assure that customers got dry helium. This problem with tank cars was relieved when permission was received to forego hydrostatic testing of railway tank cars for a 10-year period. Cylinders were tested with water, but drying them was not too troublesome. When we went to a higher purity product from the government plants, free hydrogen was found in helium from each of the government plants. Despite the claims of textbooks on the subject that free hydrogen was never present in natural gas, that was where we found the troublesome material. Minute amounts of hydrogen in helium caused objectionable porosity in welds--particularly on aluminum. The hydrogen was eliminated by passing crude helium through a catalyst bed where it was converted to water.

Although argon has become the major inert gas used in welding, helium is still used for special applications. It has been used to weld aluminum, stainless steel, copper, magnesium, titanium, zirconium, and other metals that are subject to contamination from the atmosphere during welding.

Large quantities of helium were used from the late 1950s into the 1960s as a protective atmosphere in the industrial production of titanium, zirconium, and hafnium, in a process developed by the Bureau of Mines. In the molten state, the metals have a high affinity for all gases except the noble gases--helium, neon, argon, krypton, and zenon. In fact, titanium has been used at high temperatures as a "getter" to purify gases. Automatic welding processes are used for welding stainless steel tubing from sheet material and for the mass production of aluminum beer barrels.

During the 1950s American Cyclops built a special-purpose metalworking plant near Pittsburgh in which the entire building was pressurized with an inert gas. Men entered the building through air locks and worked in spacesuit-type clothing. Most of the equipment was small, but the reactive metals could be melted, forged, cast, and rolled without contamination. Argon was actually used as the inert atmosphere, but helium was the first choice because it is easy to purify--it being only necessary to remove impurities from helium by working on those impurities. With argon, the entire gas mixture had to be liquefied and the impurities then removed.

Other metallurgical uses of helium include bubbling a stream of helium through molten metal to remove objectionable dissolved gases, as a protective atmosphere and heat transfer medium in furnaces for growing crystals such as silicon and germanium for producing transistors, as a purge for cooling vacuum furnaces, as a protective atmosphere in depositing thin metal films, and for processing nuclear fuels.

One last item in this category. In 1968, a helium time column was erected at Amarillo, Texas, to commemorate the 100th anniversary of the discovery of helium. The column was made of stainless steel joined by helium-shielded arc welding. Compartments to be opened at intervals of 25, 50, 100, and 1,000 years are filled with various objects that are likely to be of future interest. Each compartment is pressurized with helium. As I remember, the 1,000-year column contains some crabgrass seed. If any of you are concerned about the possible loss of crabgrass, don't worry.

It will be around for a long time. The original documents of the Declaration of Independence and the Constitution on display at the Archives Building are sealed in helium.

Aerodynamics

Large and small wind tunnels using helium are employed for model testing. Tunnels for super and hypersonic speeds are small in diameter, and work has been done up to 50 times the speed of sound. One report mentions study in the range of Mach 1 to 100. Helium is used because when it is expanded at room temperatures from high pressure to the low pressure required for achieving high velocities, the helium warms instead of cools. Other gases--air for example--would tend to cool and condense. When cooled to about 30°K before expansion, helium behaves about the way other gases do. This low inversion point explains why the early experimenters had trouble liquefying helium.

Shock tunnels made from converted large-bore naval guns have used helium compressed up to 100,000 psi for model testing. Work has also been done on transpiration cooling of aerodynamic surfaces with helium.

Missiles and Rockets

After World War II, the United States brought German rocket experts from the V-2 project at Peenemunde to this country for the purpose of developing a new missile capability. Little attention was paid to developing a space capability until the Russians put their Sputnik satellite into earth orbit in October 1957. That was followed a month later by the flight of their space dog Laika. Pictures released by Russia at the time showed the interior of the dog-carrying capsule where containers with the label "helium" could be seen. No mention was made of how the helium was used although they may have been testing a helium-oxygen breathing mixture.

The Russian success sparked the drive in this country to develop the capability to function in space. Our first satellites and earth orbit missions were accomplished with vehicles originally developed for military purposes. Our first effort to put a grapefruit-sized satellite in orbit failed when the Navy's Vanguard exploded on the launch pad. Some of those early vehicles used nitrogen for fuel and oxidizer pressurization, and I remember a conversation with the Air Force on the folly of using nitrogen when helium was available.

16

Adaptation of military vehicles was successful, and we remember launches of Atlas-Centaur and Titan. Atlas probably gives the best example of helium use at that time. Atlas was designed as a ballistic missile for delivering a warhead to a great distance. Helium was used not only for welding the structured components, but for pressurizing the propellants. Upon ignition, helium pressurization expelled the liquid propellants from the tanks through the engines. Residual helium pressure inflated the thin-walled metal tanks to give the vehicle structural rigidity as it flew ballistically to the end of its flight. Pressurizing with helium instead of nitrogen reduced the missile weight by 2 1/2 tons and is said to have increased the range of Atlas by 500 miles. More helium was used for engine testing than for any other purpose. Each engine was test fired several times before being placed on a vehicle for flight. This practice continued on through the Saturn-Apollo program.

Design changes in the Saturn-Apollo vehicles avoided filling those larger propellant tanks with helium as in Atlas. Before ignition, a blanket of helium gas was placed above the liquid oxygen. Some of the liquid oxygen was vaporized and brought in above the helium lens to provide pressurization. This feature separated the warm pressuring gas from the cold liquid and prevented loss of pressure from condensation of the warmer gas in the cold liquid. Weight saving in Saturn's first stage is 1,800 pounds. Because of the large volumes of liquid oxygen in the tanks, some means was required to prevent localized heating of the liquid and the formation of bubbles which could erupt like a geyser. A small stream of helium was bubbled through the oxygen to prevent geysering. Early versions of the Saturn second stage were provided with a honeycomb insulation that was purged with helium to minimize frost buildup and weight gain before launch.

Use of liquid hydrogen in second and third stages posed some special problems because of the low temperature of the hydrogen and its flammability. Helium was used to purge and pressurize hydrogen lines, valves, and fuel tanks. Second and third stage engines were precooled with helium before launch to provide a cold, purged engine chamber prior to subsequent ignition of those stages.

The third stage auxiliary propulsion system, the Apollo spacecraft and the lunar module propulsion systems all used helium-pressurized hypergolic propellants. That is, the fuel and oxidizer ignited spontaneously upon coming in contact with each other. This arrangement provided capability for multiple restarts. Helium on board the lunar module was stored as a cold, dense gas. Work has been done to eliminate helium purging and pressurization from systems using liquid hydrogen. But as far as I know, the decision has been in favor of using helium for safety and reliability. Helium has served the space program well but no less well than it has served nuclear energy.

Nuclear Reactors

Some time after World War II there was a cryptic statement from the head of the Manhattan Project that the atomic bomb would have been impossible without helium. I have heard reports that the diffusion cells for uranium enrichment were purged with helium after being opened for repair because helium could be pumped faster than air to evacuate them and get back on stream.

As early as 1942, helium was considered as a coolant for reactors. Helium is transparent to radiation, has a high heat transfer rate and can be readily purified to remove contaminants. The first helium-cooled reactor for power is generation went on stream in Germany in 1968. Its capacity 15 megawatts electrical (MWe). The year before, a prototype 40-megawatt reactor went on stream at Peach Bottom, Pennsylvania. It operated until 1973. These first reactors used to produce electricity were preceded by installations that checked out concepts but produced no power. Many gas-cooled reactors have been built in Europe, but most of them use carbon dioxide as coolant. There were reports a few years ago that some of these might be converted to helium cooling.

Helium-cooled reactors heat helium in their core to about 1,300°F. The heated helium is passed through heat exchangers to produce 1,000° steam and then returns to the reactor core to pick up more heat. Conventional steam turbines are used to turn the electrical generators. The helium-cooled reactor offers the possibility of eliminating the steam cycle and removing the heat energy from the helium by passing it directly through a suitable turbine which would rotate the electrical generator.

Where pellet-type fuel rods are used in reactors, the space between pellets is filled with helium to improve heat transfer. Some fuels are contained in cans and addition of helium to these cans before sealing not only improves heat transfer but provides a built-in means of checking for leaks. The canned fuel elements can be passed through a helium leak detector, and if no helium is found, the seal is tight. Welding and the use of helium for leak detection in reactor components has already been mentioned.

I am omitting more than a mention of the breeder reactor and the generation of power by nuclear fusion. These uses are in the future although considerable work has been done on each of these two concepts.

Thus far, I have talked about the use of helium in thermometry, as a lifting gas, in diving, in leak detection, in chromatography, in welding, in metallurgy, in wind tunnels, in missiles and space vehicles, in nuclear reactors, and other uses. Finally, we come to what is now the largest use category.

Cryogenics and Superconductivity

The term cryogenics comes from the Greek word cryo meaning "icy cold," and it refers to scientific and engineering work done at low temperatures--generally accepted to be below about -150°C. The first attempt to liquefy helium in 1895 ended in failure; so did other attempts in 1901, 1903, and 1905. By 1905, all the so-called permanent gases had been liquefied except helium. Hydrogen that boils at about -253°C was used for cooling helium in those first attempts to liquefy it. In 1908 Kammerlingh Onnes successfully liquefied helium and found that helium would liquefy only if precooled to a low enough temperature before it was expanded to a lower pressure. The boiling point of helium was found to be about -269°C or only a few degrees above so-called absolute zero. A whole new world was opened to exploration. Onnes' supply of helium was limited. He had about 13 cubic feet that had been obtained by heating monazite sand. Later he got more from that source. Work then began to learn more about properties of the new liquid and to study the behavior of some materials in this range of temperatures. By 1911 Onnes had discovered superconductivity in mercury. He found that when he cooled mercury to liquid helium temperature all electrical resistance disappeared. Within two years, he knew tin and lead were also superconductors, but to his dismay he also learned that a magnetic field destroyed the superconducting property. The idea of using superconductors lay dormant until about 1960.

Helium received attention in laboratories in Europe and in the United States. Liquefaction was a tedious and potentially hazardous undertaking because of the need to use hydrogen for precooling; however, much was learned about liquefying helium.

When helium was solidified, it was found to be a superfluid that would run uphill. At a very low temperature, solid helium became a perfect heat conductor.

The availability of helium for cryogenic work took a leap forward after Peter Kapitza in Europe in 1934 and

Samuel Collins at MIT in 1939 each demonstrated helium liquefiers, using expansion engines. No hydrogen was needed. Before long anyone could buy a helium liquefier or refrigerator. With that development, helium began moving out of the low temperature laboratory.

Today, large commercial liquefiers are in operation and the liquid is trucked across country as readily as if it were milk. We have already mentioned helium cooling of rocket engines. Helium has been used in space simulators for duplicating the near absolute temperature of outer space for testing full-size spacecraft. The liquid has been used in bubble chambers to study the behavior of nuclear fragments. Some tissue cultures--notably in cancer research--have been preserved in it. In the late 1950s the National Bureau of Standards had a project to study frozen free radicals by freezing them in liquid helium. The idea was to extend their life beyond the normal millisecond range. Theoretically, the controlled release of their energy would give a very high specific impulse. As I recall, work was discontinued when results were disappointing.

The first ground stations for amplifying signals from communications satellites used liquid helium-cooled masers. Later, these were replaced with gaseous helium-cooled parametric amplifiers. Miniature helium refrigerators are used in aircraft to cool infrared sensors.

One of the biggest cooling jobs helium is called upon to do is with superconducting magnets. During the 1960s much effort was put into developing new superconducting alloys. Fabrication for practical applications posed major problems. Enough of those problems were solved so that several high field superconducting magnets are in use. One large, superconducting magnet went into operation at Argonne National Laboratory in 1969. It produces a magnetic field intensity of 18,000 gauss. Its cost was $2.4 million. Power costs to operate the magnet for 10 years were expected to be about $400,000. Power costs for a conventional magnet to do the same job over the same time period were estimated to be $4 million. Other high field superconducting magnets have been built. One was delivered in 1969 to a German firm for use in the magnetohydrodynamic (MHD) generation of power. I read the other day that the United States flew a large superconducting magnet to Moscow last summer for use in their MHD project. MHD power generation involves passing on ionized gas obtained by burning coal, for example, through an intense magnetic field where electrons in the ionized gas are captured and fed into the electrical distribution system.

Superconducting underground power transmission lines are
in the planning and development stage. The current carrying
capacity of cable systems can be increased by a factor of 20
over conventional systems. A superconducting electric motor
was built and used in England a few years ago. Its job was
to pump water from a mine. You may recall Newcomens' first
steam engine was used for the same purpose.

Other applications of superconducting devices include
portable power plants, shipboard motors, and computers. The
cryotron, a superconducting device, was developed in 1956.
Its application to computers has been considered since that
time.

One final word about helium in cryogenics--a great deal
of helium is used in carrying out research that may result
in the improvement of present devices and the discovery of
new ones.

There is at least one area of helium use in cryogenics
that I have not mentioned. That is the use of helium-3. I
have been talking about helium with the mass of 4. The
lighter isotope helium-3 has been used as a gas in neutron
detection devices. Mixtures of liquid helium 3-4 have
special properties. Their use has resulted in achieving
temperatures within a millidegree of absolute zero.

I referred to masers earlier, but the maser has a
cousin, the laser, which operates at warmer temperatures.
The laser is a device that provides a beam of coherent
light. That is, the beam does not scatter in a normal
manner and can be projected across great distances. There
are several kinds, some operating with solids, some with
gases. The helium-neon laser which dates from about 1961
was the first to produce a continuous beam--others at that
time could produce only short bursts of energy. Continuous
beam lasers have been relatively low in energy while beams
from pulsed lasers are powerful enough to burn steel. Beams
from helium-neon lasers were used to align the excavation
across San Francisco Bay for laying the prefabricated tubes
of the BART subway system. I read the other day that the
garment industry in this country cuts cloth for suits with
lasers while other countries still use scissors. I do not
know whether cloth-cutting lasers use helium but helium is
used in combination with other materials in recently
developed lasers.

I have drawn freely not only on my own experience and
publications but the publications of others, particularly
the continuing work of the Helium Division of the Bureau of
Mines.

DISCUSSION

CLARENCE T. KIPPS, JR., attorney, Miller & Chevalier: I
represent Northern Helex Company, which is one of the four
helium conservation contractors. Northern Helex filed suit
against the government in December 1970 for breach of
contract. The Court of Claims held that the government
breached the contract and that Northern Helex is entitled to
recover its damages. The amount of damages is still being
litigated.

One of the issues extensively tried by the government
and Northern Helex before Trial Judge Spector in 1973 was
whether it is in the public interest to conserve helium.
Trial Judge Spector's December 3, 1974, report contains an
objective and comprehensive discussion of the helium
conservation program, the cancellation of the contracts, and
why it is in the national interest to conserve helium.
These conservation issues were not considered by the court
in its October 1975 decision because the court rejected the
government's termination argument on a legal ground without
reaching the merits of helium conservation. Trial Judge
Spector's report provides the Committee an objective data
base upon which to evaluate the assertions now being made
against further helium conservation.

It is strange to find responsible government officials
acquiescing in and indeed encouraging the waste of the bulk
of this nation's proven helium reserves. The reason for
such an unusual course of action is that the administration
in 1970 made a very bad decision when it decided to
discontinue further helium conservation for budgetary
reasons. The government is still attempting to defend that
decision on the basis that the Helium Act authorized
conservation of helium for only "essential government
activities" and that the government now has stored enough
helium for those activities. Thus, we find that this
country's helium policy is now being dictated by a bad 1970
Office of Management and Budget decision and the litigation
between the government and private parties.

Helium is too important to the scientific and
technological growth of the United States for our leaders to
continue a total "hands off" policy because of the effect
any action might have on the litigation. In the first
place, the government's legal position is wrong. It is
going to lose the litigation in the Court of Claims. In the
meantime, the helium will have been wasted. In any event,
this Committee is concerned with the need to conserve helium
for national requirements and is not limited to evaluating
the helium needs for essential government activities. Thus,
this Committee can focus on and aid in establishing a helium

22

policy for the benefit of the nation without undermining the government's primary legal defense in the litigation.

With this background I now turn to the basic question posed for discussion. Should helium be conserved in the national interest? Yes. It is clear that this valuable, unique and limited natural resource should be conserved so this nation's supply of helium is stretched as far as possible into the next century.

A. Trial Judge Spector's findings of fact attest to the need to conserve helium for the benefit of the nation as a whole.

B. In 1969 when about 3 billion cubic feet (Bcf) of helium were being conserved under the nation's helium conservation program, the National Academy of Sciences' National Research Council Committee on Resources And Man in its study and recommendations was concerned about the adequacy of the program. The Academy stated:

> 3. That the present Helium Conservation Program of the Department of the Interior be reevaluated. Helium is unique in its combination of unusual properties and critical uses. It is essential for cryogenics, superconductivity, cooling of nuclear reactors, exploration of the seabed, and the space program. According to available estimates it is in short supply, yet it continues to be wasted in the combustion of natural gases. Its recovery from these gases and conservation for the future is feasible and is already being done on a limited scale. The Helium Conservation Program should be carefully reevaluated to determine if it can meet helium needs beyond the early part of the 21st century. If such evaluation leaves any question at all about the adequacy of the program, the program should be extended without delay to apply to lower concentrations of helium and more natural gas fields. (pp. 10-11)

C. Today, the government is saving less than 200 MMcf of helium out of the 8 Bcf now being removed annually in natural gas produced for fuel. Of this 8 Bcf, about 1 Bcf is actually separated from the gas stream (because of the integration of the privately-owned helium plants with other facilities) and then wasted.

D. Except for the helium in storage at the Cliffside reservoir, 85 percent of this nation's proven helium reserves are in the natural gas now being produced for fuel from fields located in Texas, Oklahoma, and Kansas. Unless extracted and saved from the gas on the way to market, the helium is wasted into the atmosphere. All of these gas fields will be depleted before 1955. All of this nation's presently known gas fields containing helium will be exhausted by the year 2000.

E.   Government owned and privately owned helium extraction plants now in place are capable of extracting about 4 Bcf of helium per year which could and should be stored for the nation's long-range helium needs. The government's Cliffside reservoir can store all of this helium at a modest storage cost.

F.   The helium program which was the subject of the National Academy's earlier study was correctly described by the Department of Interior as it was then (and now should be) as follows:

The helium conservation program is not a stockpiling program aimed at assuring an adequate supply of helium for some predetermined uses and for some predetermined period of time. It is a conservation program aimed at curtailing the wastage of valuable natural resource in order that the resource will be available to future Americans for whatever purpose and at whatever time it is needed.... (U.S. House of Representatives, 1964).

Under Secretary of Interior Bennett, in supporting the legislation establishing the program, most appropriately stated:

Perhaps the known resources will be capable of meeting our needs to 1995 or the year 2000, instead of 1985 as predicted in our estimates. At some time in the future, we will surely need the helium that is being wasted today. (U.S. House of Representatives, 1960).

G.   It is uncontroverted that large quantities of helium will be neded in the future. All of the forecasted requirements for helium show that they exceed our suppplies and that eventually we must rely on extraction of helium from the atmosphere. The Helium Act gives the Secretary of the Interior authority to control the use of helium including the price to be charged when it is sold. Therefore, from an economic standpoint, the government can set its sales price at whatever it wants to assure that it does not lose money on conserving the helium.

H.   In his April 10, 1975, covering letter, the Administrator of the Energy Research and Development

Administration highlighted the essential thrust of his Report to the President and the Congress entitled The Energy Related Application of Helium. He pointed out that: (1) an analysis of energy technologies under development indicates that "substantial amounts of helium will be required in the future--largely in the next century"; (2) "virtually all domestic helium, which is now contained plentifully in natural gas, is expected to be depleted by the end of the century-- before new technologies require it" and (3) this nation will have to depend on "foreign supplies of helium" and extraction of helium from "the atmosphere--a source from which helium is presently 400 times more expensive than from currently available sources."

I. In passing Senate Resolution 253 expressing the sense of the Senate on helium conservation the Senate Committee Report pointedly states:

...It is short-sighted in the extreme to foreclose future development of technologies which are bound to determine to a large extent whether or not the United States can meet its future energy requirements. When the risks to national growth and security are assessed, the Committee can only conclude that further delay in removing whatever obstacles stand in the way of helium conservation is totally unjustifiable. (U.S. Senate, 1976).

I am pleased that the Academy has been asked to evaluate the present federal program for the conservation of helium and its adequacy to meet future needs. If I can provide any additional data to assist in this important effort, I want to do so.

F. CLAYTON NICHOLSON, Northern Natural Gas Company: I am a member of the Committee and I would like to direct my inquiry to Harold Lipper, who gave an excellent presentation of the developments of the past. It seemed to me that his remarks showed the really dynamic nature of the efforts being made in the use of helium. Would Mr. Lipper anticipate that this type of activity would continue now that we know so much about the properties of helium?

LIPPER: I think that the future probably is very strong and bright. The Midwest Research Institute (1977) report that was done for the Bureau recently indicated a strong potential in the use of helium in cryogenics and cryogenic-related research.

CHARLES W. HOWE, University of Colorado: How many of the accomplishments that you have recited in your very interesting talk might have been accomplished with the use of substitute resources? Could you comment on that to give some of us who are not from the hard sciences, so called, a little insight into how many of these steps might have been taken either at the time or eventually with substitute resources?

LIPPER: Well that really opens a can of worms, but I can do something with it. One of the things I mentioned was the use of an inert gas for welding because of the physical properties. People wanted to use argon; but when they had a job to do in a hurry they found that the argon they wanted to use, because of other circumstances, was wet. Well, it took a little while to dry the argon. But helium was dry, and it was available to do the job at the time the job needed to be done. In the space program people found that they could do some pressurization jobs with other materials. But the one area in which there definitely is no substitute for helium is in the field of cryogenics. You can only get down to about 20°K with hydrogen, and with ordinary liquid helium we're down to 4° above absolute zero. Some of the work that I've seen the papers on goes down to within a minimal degree of absolute zero. So in these very low temperature regions it would seem that you have to have helium. Until someone comes up with a reliable superconductor with a transition temperature above that of liquid helium and with the capacity to absorb shocks, any superconducting systems will have to have helium if they are to work.

LEWIS H. NOSANOW, National Science Foundation: I'm head of the Condensed Matter Sciences Section, Division of Materials Research, National Science Foundation, which is responsible for the support of a large fraction of the research that is done in the areas of low temperature, solid state physics, and solid state chemistry. We are supporting a great deal of research on superconductivity and solid state physics, and it appears to me that there are substantial breakthroughs yet to come, all of which will require temperatures that can be achieved only by liquid helium-4 or liquid helium-3. We get continual reports of new research and many proposals. I think that it is one of the vital areas of research and on the forefront of science, a supertechnology 10, 20, 30, 40 years from now.

DRAKE: Any other questions or comments from the general audience? I will ask Mr. Lave if he would like to give us a preview of his comments scheduled for tommorow.

LESTER LAVE, Carnegie-Mellon University: Let me start off with the observation that extracting helium from natural gas is probably one of the best examples of the second law of thermodynamics I know about. When you start off you can have helium-rich natural gas that has concentrations as high as 8 percent helium, and you're talking about roughly 80,000 parts per million (ppm); whereas, in the atmosphere you have about 5 parts per million. It is quite clear that as soon as you have burned the natural gas helium goes from a concentration of 80,000 to 5, in the course of which entropy has increased enormously. Once that's done, the only way of retrieving what you've just wasted is to put an enormous amount of energy back into concentrating the helium.

Clearly an illustration of this sort, which is so simple and so compelling, must be able to convince even economists. But alas we are not convinced. Economists must be the last of the allegedly intelligent people who don't believe in the laws of thermodynamics. I'll go into more of that tomorrow. The principal reason why we don't believe in them has to do with the social rate of discount, that is, the fact that a dollar tomorrow is not worth as much as a dollar today. It all has to do with something that we academicians talk about a lot--deferring gratification. Just how much gratification are we willing to defer for the future either for ourselves, our children, or for general future generations? That leads to two questions that I'm going to pose in detail tomorrow. How generous do we wish to be to future generations? That is, how much do we wish to reduce our current consumption so that they might have more to consume? And the second question: However generous we decide to be, how are we going to invest these resources that we're going to save now. That's a very complicated question. We could decide to be very generous and be very stupid. I think about the illustration of our great grandfathers deciding in 1870 that they were going to be very kind to us in filling up warehouses with horse collars so that we would never need to produce them again. And so people have to not only decide how generous they're going to be to future generations, but also what is the intelligent investment to make?

That gets to the question we heard about a minute ago: Is it really in the social interest to conserve helium? Certainly that is an interesting question, but it is quite different from a second question which is: If one should decide that it is in the social interest to conserve helium, what should be the role of government? There one can start off as a doctrinaire person looking at government and

27

saying, on one of the two extremes, either that the role of
government is to do all in the economy--that is, one could
talk about a government controlled society--or one could go
all the way to the right and say that the role of government
is to provide us with a police force and national defense
and that's all; or, one could say that the government ought
to intervene in the market, selectively, depending on
special cases. Economists always like to talk about special
cases, and they have a whole list of special cases where the
government might want to intervene in the market. These
include general cases of market failure such as, for
example, lack of competition or the presence of collusion in
an industry; some economies of scale that the market can't
really handle in a competitive fashion very well; something
that we like to talk about as public goods, that is, in
particular, either the environment or national defense; and
then the fact that government might have some special
knowledge about the present or the future that other people
don't have. If one has an affirmative answer to the
question of should helium be conserved, and then goes on to
ask if the government ought to be intermixed in this
business of conserving helium, one could try and go down
this checklist to see whether there are important reasons
why the government should be there. I'll be pleased to do
that in a fairly provocative way tomorrow. But in the end,
as an economist, I'm drawn to the last question that Charles
Howe asked, namely, to what extent is helium titled in the
current Act a unique, nonrenewable resource? It seems to me
that although the government, at least the Congress, has
frequently declared something to be such and such that we
have in fact often found after looking back on it that it
wasn't that.

DRAKE: I think we are making progress. The economists are
now mentioning at least the second law of thermodynamics.
Mr. Laverick, would you like to give us a preview of your
material scheduled for tomorrow?

CHARLES LAVERICK, consultant: I was asked to talk about
future requirements and uncertainties. The policymakers and
economists point out that most of the future uses of helium
seem to be too speculative to worry about, and, of course,
that time ends at around 2000 A.D. So it seemed to me that
one of the things I could do was to tell you roughly where
we stand today with the technology that didn't exist in
1960, show you how remarkable the progress has been, and
then point out where we hope to be tomorrow. [There
followed a summary review of the Laverick presentation which
appears later in this document.]

     In terms of uncertainties, we have to ask ourselves:
What is the world of the future going to be like? The
economists say that things are always going to get better:

we're not so well off today as we will be tomorrow. Maybe
it's true, maybe it's not. Maybe the world's resources are
infinite in the terms of our society, maybe not. Who knows?

The environmental impact statement took the view that
we've got 80 years supply if we think of government agency use;
we've got 6 months supply if we think of national use. So
why bother? And I say that if you have a reasonable amount
of helium that will last you 20 or 30 years at some rate
like 2 billion cubic feet a year, you can do a lot. These
are some of the things I would hope to talk about.

ROGER DERBY, Department of Energy: I don't think anyone has
mentioned this evening which other countries have helium. I
suspect that the United States is one of the few countries
that is rich in the supply of helium. I wonder if there's
an impact on our balance of payments or in our diplomatic
relations with other countries?

NICHOLSON: I think I can address myself to that question.
The largest plant outside the United States operating today
is in Poland. This plant is processing gas from a field in
Europe that is high in inerts. In the preparation of that
gas for fuel markets, they are extracting the helium. That
plant is operational and provides about 150-200 million feet
a year for the European market. Canada had a small plant
that operated for a number of years but that has been shut
down, because, I think, it's not competitive in the cost
structure today. There's a bit of helium being produced in
an ammonia plant in Holland, where the inert stream from an
ammonia plant has a concentration of helium that enables a
small unit to be installed there. There are other potential
plants outside the United States, but none of them are
active projects at the moment.

DAY II

ROBERT M. DRAKE, Jr.: We now have learned quite a bit about helium. The second lightest element next to hydrogen is a colorless, odorless, monatomic gas, which although chemically nonreactive seems to be reactive in the sense of societal and legal activities. We learned too that helium is abundant in nature, second again only to hydrogen, and is formed by radioactive decay of uranium and thorium in the crustal earth. And though it is abundant on earth, the gas is widely diffused in low concentrations, mostly in the atmosphere, where the equilibrium concentration is about 5 ppm. In contrast to that low concentration in the atmosphere, helium is contained with natural gas in much higher concentrations, from which separation extraction is of current interest technically and economically and perhaps of even greater interest in the future.

Currently helium is used medically for respiratory therapy, for welding, for divers' breathing apparatus, for toy balloons and weather balloons, gas-cooled nuclear reactors, leak-detecting systems, inert atmospheres, inert purge and pressurization systems, heat transfer applications, lighter-than-air lifting systems, and wind tunnels. It is used in cryogenic systems in support of superconducting magnets, superconducting conductors, superconducting rotating machinery, and superconducting transmission systems. All such current uses apparently are more than adequately met by present helium inventories and supply. In the future, perceived uses of helium are predicted to require increasingly larger supplies of the gas, assuming that the technologies upon which that perception is based do develop and become widespread commercially.

In this context, we heard mentioned gas-cooled nuclear reactors, superconducting electric machinery, superconducting power transmission, MHD and controlled thermonuclear fusion. It seems that the present irrationality in regard to helium arises from an earlier imperfect bit of legislation which, however noble its interest and intent, seems perfectly devised to invite the inevitable disharmony and litigation that has occurred.

We now will move further into the exploration of these key issues in regard to helium, with presentations on the

cost of helium in relation to the present supply, the future
requirements and uncertainties in regard to the resource,
and finally, the ultimate bottom line, the management of
helium resources.

An Overview:
SHOULD THE GOVERNMENT BE HELIUM'S FUTURE

Lester B. Lave
Professor and Head
Department of Economics
Carnegie-Mellon University

Introduction: Economics Versus Entropy

The second law of thermodynamics is well illustrated by the extraction of helium from natural gas. If the helium is not extracted before combustion, it diffuses into the atmosphere, going from 5,000 parts per million (ppm) to 5 ppm. Perhaps 1,000 times more energy is required to extract the helium once the natural gas has been burned.

Surely this illustration of the second law is plain enough for anyone to grasp. Surely even economists must agree that it is better to extract helium from natural gas now than from the atmosphere eventually. But economists remain unconvinced. You might have expected better from a discipline that predicted ruin a century before the formulation of the second law. Malthus' predictions of a population continually outstripping the food supply and constantly having to look to war and pestilence for salvation made economics, not thermodynamics, the "dismal science."

The economic explanation, unfortunately, is more complicated than the story about helium and entropy. It requires an examination of current and future supplies of helium, both in terms of the resource base and extraction technologies. It requires an examination of current and future demands for helium. In addition, there is the question of the social rate of discount, the role of the private sector, and the extent of externalities. In order to keep from confusing themselves, economists distinguish between real and monetary phenomena. These two topics are the next headings, with a summary section on government's role that follows.

Real Phenomena

Is helium a scarce, irreplaceable resource? Obviously,
the answer is no, whatever Congressional preambles say.
Helium is the second most common element in the universe
with vast quantities on earth. While it is hard to term any
event inconceivable, running out of helium is as close to
inconceivable as I can come.

Furthermore, in virtually all current commercial uses,
helium could be replaced by some other substance. Perhaps
the only irreplaceable use for helium is in superconduction,
but note that there is no commercial application in
superconduction at present.

If helium is not scarce or irreplaceable, what is the
issue? The nugget of the answer involves choosing among
three actions: (1) Use natural gas and dissipate the helium
into the atmosphere. (2) Use natural gas and separate the
helium (at a considerable cost) for use sometime in the
future. (3) Or not use the natural gas containing helium
until either we are sure that we do not want the helium or
are sure that we do want it and are willing to pay the cost
of separation. The first action is the cheapest now but may
lead to high costs in the future if the only way to obtain
helium is from the atmosphere. Actions (2) and (3) are much
more expensive.

There is estimated to be 777 Bcf of helium in U.S.
natural gas. The cost of extraction depends on the
concentration of helium, with some gas fields being
extraordinarily rich in helium (more than 80,000 ppm) and
others being much poorer (less than 650 ppm). Current
extraction costs would run from perhaps $10 for very rich
fields to perhaps $130 for poor fields (per 1,000 cubic
feet--Mcf). There are also other sources of helium. There
are probably underground pockets, although no technology
exists for finding them other than random drilling.
Finally, the atmosphere is virtually an infinite source of
helium, but at a concentration of only 5 ppm.

Figure 1 shows a supply curve for helium with cubic feet
on the horizontal axis, and dollars per cubic feet on the
vertical axis. The supply curve looks almost discontinuous
in the sense that there is a rising curve as we think of
going to leaner and leaner natural gas supplies, that is
natural gas that has less and less helium in it. Then it
takes an abrupt jump when we go from natural gas supplies to
liquefaction of the atmosphere, with a very considerable
increase in price. So the supply curve doesn't appear to be
continuous. Yet even if we take a look at the lower portion
of the curve, we can see that the costs rise appreciably
depending on how much we want to pull out of it. We can

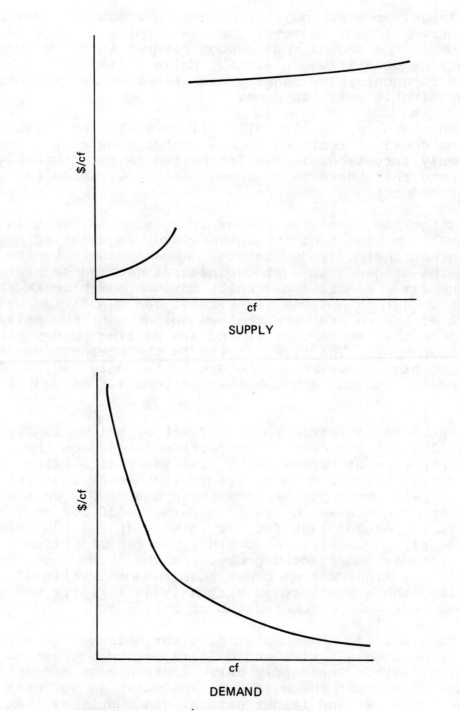

SUPPLY

DEMAND

FIGURE 1 HELIUM SUPPLY AND DEMAND

make a judgment as a society to pull all of the helium out
of natural gas as long as the natural gas contains, say, 10
ppm or more. If we did that, then we would be talking about
costs that would be quite high, and a curve that was not
quite so discontinuous.

At the bottom of Figure 1 I've taken a look at some of
the essential factors in helium demand. First of all, one
can begin with some of the essential uses, those that depend
on the cryogenic properties of helium. Here, the number was
mentioned last night of helium increasing the costs of, for
example, fusion by 10 percent if we were to have to go to
atmospheric extraction. We had a discussion with ERDA a
while ago in which Harvey Brooks remarked that one had to be
careful of the economists since it wasn't clear that anybody
should believe their projections. I had to remark that as
an economist, one ought to be very clear about the
technological projections, since it wasn't clear anybody
should believe them either. Here especially, when somebody
talks to me about a 10 percent increase in the price of
fusion, I must say I think that's like counting the angels
on the head of a pin. If you have a technology that might
or might not come about, that might or might not have
commercial possibilities if it does exist, and that might or
might not be cheap if it does exist, and then you say that
if you have to extract the helium from the atmosphere it
will raise the cost by 10 percent, I frankly don't know what
you're talking about--that's my prejudice.

The future cost of helium extraction depends not only on
the resources available but also on the rate of
technological change in exploration and extraction. For
example, helium might be extraction from the atmosphere
without liquefaction. One could calculate the theoretical
energies required for separation of the helium and conclude
that there need be no reason for the cost of extraction from
the atmosphere to be 1,000 times present cost.

In summary, the future supply of helium remains a
mystery. It depends on discoveries of natural gas fields,
as well as of the concentration of helium in these new
fields. ERDA's MOPPS study estimated the availability of
thousands of quads of natural gas at somewhat higher costs
than present, a large increase in estimated resources.
Helium supply also depends on new technologies of exploring
for helium directly. Most importantly, helium supply
depends on technological advances in separating helium from
the atmosphere.

The vast proportion of the current demand for helium
rests on its proporties of being light and inert. In
essentially all of these uses, other substances could be
substituted. Even in the cases where helium is unique, the

use could be reconfigured so that less helium was used and
most of that used was recycled. Thus, the demand for helium
will vary with its price; if the price of helium were
sufficiently high, little would be used. It seems likely
that prices 50 times higher than present prices, predicated
on atmosphere extraction, would reduce use to low levels,
probably consisting only of those uses where it was
irreplaceable.

In conclusion, the economic viewpoint of supply and
demand tends to paint a rosy picture: Large quantities of
helium will be available and the price will serve to
equilibrate supply and demand. There is no possibility of
running out of helium, unless government regulation
precludes the price from rising to the point where supply
and demand are balanced.

Monetary Phenomena: The Rate of Discount

We could extract helium from natural gas all over the
world in order to keep it from dissipating into the
atmosphere. We need only offer a suitable price, say $100
per Mcf to all suppliers, domestic and foreign. By
expending hundreds of millions (or even billions) of dollars
each year, we could accumulate tens of billions of cubic
feet of helium each year in order to prevent entropy from
increasing needlessly. If we did so, would our
grandchildren bless us? Would we bless our grandfathers if
they had bequeathed us warehouses of horse collars, or
preserved the rich iron and copper ores? That is, would we
be better with the Mesabi range intact today or with the
railroads, bridges, and factories that were built from its
steel?

It makes an enormous difference in economics in what
order things are done. Eating bread now and planting wheat
next year is very different from planting wheat now and
eating bread next year. A small amount of wheat seed
produces an enormous amount of flour--next year--but only a
small amount of flour now.

There are two central questions that economics poses:
(1) How generous do we want to be to future generations (how
much consumption should we forego now so that future
generations can consume more)? (2) For whatever amount of
consumption we choose to forego, what is the best way of
investing it?

When we decide to save wheat seed, instead of having a
little more bread to eat now, we are foregoing current
consumption in order to have higher consumption next year
and in subsequent years. In a complex, industrial society,

there is not quite so direct a connection between foregoing current consumption in order to have more to consume later, but the process is just as real. We save for our immediate future, for our long-term future, and for the future of our children and all subsequent generations.

But simply foregoing consumption does not increase the amount of consumption available to us in the future. That foregone consumption must be invested wisely. This investment can be in machines, tools, buildings, and conventional capital goods. Instead, it can be in research and development in order to increase the efficiency with which raw materials are transformed into the products and services we desire. Or it can even be in what economists have come to call "human capital," the amount of knowledge we have in our heads, our set of skills, and our physical condition. There are many good ways of investing in our future, but there are also many investments which are unproductive, or even pernicious. For example, our grandfathers could have invested in the Mesabi range, the domestic steel industry, research and development in the concentration of lean ores, basic oxygen furnaces, and steel technology, or the training of geologists, metallurgists, and the health of workers.

What Should We Invest In?

Had our grandfathers carefully invested in horse collars or wrought iron and saved them for us, we would not be thankful. We would be richer if they had invested in steel or transportation companies. Investments in natural resources have not done as well as investments in corporations. With a real rate of return of 7 percent per year, an investment doubles about every 10 years. Thus, a dollar invested in 1927 would be worth about 30 1927 dollars in 1977; after accounting for inflation, this would amount to about 106 1977 dollars. This is a spectacularly good investment by any criteria. No mineral would have increased so much in value over this period; indeed, between 1870 and 1960, all major minerals decreased in price (in real terms).

The lesson is that investments in concerns managed by people have yielded much higher rates of return than investments in natural resources. No one can tell whether this will continue to be true in the future, but it seems a reasonable bet.

A major attribute of the private sector is that it does not assume that all individuals are equally good investors. Instead, the market attempts to identify the individuals who are best at making investments and managing firms; these individuals are then entrusted with the savings of many people. As a result, the average rate of return in the economy is higher than if each individual had to invest his own savings.

The Social Return to Investment

The vast majority of decisions on savings and investment are made in the private sector. Savings depend on individual or corporate income and the rate of interest offered (the higher the better). Investment depends on the expected return, with the lower the interest rate, the more funds will be desired for investment. With income and expectations about the future held constant, there is an upward sloping supply curve of loanable funds (savings) and a downward sloping demand curve for loanable funds (investment) that determine an equilibrium interest rate and rate of capital formation.

Although the supply and demand for loanable funds is more complicated than this brief argument indicates, the equilibrium interest rate plays the central role in guiding savings and investment. For example, in a rapidly growing society, a high interest rate indicates a shortage of loanable funds; the high interest rate induces additional savings and rations the scarce funds so that only the most essential (profitable) projects get funded.

Government investment must be guided by the same general rules. Once they have been made comparable, government investment ought to earn the same rate of return as private investment.   There is a large and somewhat murky literature on the "social rate of discount," the interest rate that ought to be applied in evaluating government investments. When he was head of OMB, George Schultz directed that 10 percent be used as the rate to evaluate government projects. At that time, the inflation rate was about 7 percent and rising, and so one might have interpreted the directive as making the interest rate 3 percent in constant dollars.

One piece of evidence on the social rate of discount is to look back at the return that investors have earned over a substantial period of time. Evidence is available on both the return to bond holders and the return to equity holders.

Table 1 summarizes evidence on the real (after accounting for inflation) rate of return to investors from 1927 to 1976. The table shows similar rates of return in the periods 1927 to 1976 and 1947 to 1976. The return to equity holders is about 7 percent per year, while that to long-term government bond holders is about 1 percent per year. Furthermore, the rate of return has dropped considerably since 1962. Since 1962, equity holders have averaged less than 2 percent per year.

The pretax rate of return for manufacturing corporations is higher than these rates. However, different rates of return are earned by people for various investments. From these data, it seems likely that investors can expect to earn perhaps 3 percent per year, after accounting for inflation. This return is broadly consistent with the OMB dictum of using a 10 percent rate of return, including inflation.

Should the Government Have Anything to Do with Helium Supply?

The answer is probably no. There are many nonrenewable resources, such as helium. Why is this one singled out for government action? Should all natural resources be controlled by the government? This last question probably requires a philosophical answer, depending on whether one thinks that the marketplace does a better job of allocating resources than the government. I believe that it does and can cite much evidence supporting that, although I realize that it is now fashionable to believe that the federal government can solve all problems and that the marketplace is to be distrusted and given as little power as possible. Note that even if one believes that the government should control all natural resources, it does not follow that we should not be burning the natural gas and dissipating the helium--that depends on the rate of social discount.

But suppose that one believes that the current policy of the United States is correct, that government should not control all natural resources. Is helium an exception? It is not essential to national defense. Its collection or dissipation poses no major health or environmental problems. If helium is going to be highly valued in the future, the marketplace is perfectly capable of storing it. When was the last time you drank Scotch that had been distilled in 1977 or a fresh bottle of Lafite Rothschild? You will pay a good deal for property with shale oil or tar sands, even though it will be decades or even a century before the energy there will be marketable. I have little reason to believe that the government is better at forecasting the future than the private market.

But there are many important uses of helium in the future: superconducting transmission of electricity, superconducting magnets for fusion, or gas-cooled reactors. I am suspicious of these examples for three reasons. (1) The calculated price of fusion or gas-cooled reactors would rise by only 5 to 10 percent if the helium had to be gotten from the atmosphere at $3,000 per Mcf. Given our

TABLE 1 Real rates of return (in percent per year) of common stocks, long term government bonds, and treasury bills during selected holding periods.

Holding Period	Common Stocks	Long Term Government Bonds	U.S. Treasury Bills
Through WWII			
1927-31	-1.4	6.1	6.8
32-36	23.5	8.5	1.1
37-41	-9.3	1.6	-2.0
42-46	10.2	-3.1	-6.2
1927-46	5.0	3.2	-0.2
Post WWII			
1947-51	11.9	-3.6	-3.2
52-56	19.2	0.1	0.8
57-61	10.9	0.8	0.8
62-66	3.8	1.3	1.7
67-71	3.7	-2.7	0.8
72-76	-2.3	-0.4	-1.2
1947-76	7.6	-0.8	-0.1
1927-76	6.6	+0.8	-0.1

Source: Ibbotson and Sinquefield, 1977.

uncertainty about the technological possibilities of these technologies, as well as about their costs, the cost of helium is in the noise. (2) We do not really know what the cost of helium extraction will be in 30 years. Helium can be gotten underground other than in combination with natural gas; cheap technologies for extracting it might be developed. More efficient ways might be found of extracting helium from the atmosphere; the costs could be reduced dramatically. (3) Substitutes for helium in many applications are likely to be found. Is helium so unique that it really will be required in all these uses? It seems doubtful.

In addition, we are not behaving now as if helium were a unique resource. We are using helium for many purposes, in few of which is helium necessary. Thus, almost a billion cubic feet of helium is "wasted" each year in commercial uses. If helium were so precious and unique, why should any of it be used for these low-level purposes?

I see no reasons to believe that helium is unique and no reasons why the government should be engaged in supplying helium. The private sector is perfectly capable of separating helium from natural gas and supplying all of the current demand. Insofar as it appears that helium prices will rise in the future, companies will be motivated to store helium. Certainly companies are unlikely to store helium that they do not expect to sell for 50 years; but the future is murky, and it seems doubtful that anyone can forecast 50 years into the future with assurance.

Is Helium Storage a Good Investment?

If investors can expect to realize approximately a 3 percent (real) rate of return on investment in general, is helium a good investment? There is no way one can be sure of the future price of helium or of technological change in helium extraction. If, by 2020, the atmosphere were the principal source of helium, it is likely that helium would be substantially more expensive than at present. At present, extraction from the atmosphere is more expensive than extraction from natural gas. Even if technological change in helium extraction were to proceed at 2 percent per year, the cost would be upward of 20 times more expensive. If so, helium stored now would earn almost a 7 percent real rate of return, compared to past experience. It would be a high return either for government or a private investor. Thus, it appears that helium is indeed a good investment. But this does not imply that the government ought to be the investor.

Can Any Case Be Made for Government Storage?

As argued above, helium is a good investment, promising to earn almost a 7 percent real return. Thus, it is an attractive investment for either the private sector or government. But if the private sector fails to make the investment, should the government do it?

It is possible that private investors are scared off because of the possibility of government price or other regulation that would make the project unprofitable. If so, government action to foreclose this possibility would be preferable to government investment.

I see no evidence that the private sector is unwilling or unable to make long-term investments of this sort, at profitable rates of return. There is great uncertainty regarding the future demand for helium. Will advanced energy technologies demand vast amounts of helium? What will happen to the technologies extracting helium from the atmosphere? Will cheaper supplies of helium be discovered? These uncertainties tend to make the investment less attractive. There is, of course, the opposite possibility that the demand for helium could skyrocket because of new technologies that make more extensive use of it.

While the uncertainties make the investment less attractive for the private sector, they also make it less attractive for the public sector. These uncertainties are just as important to the rate of return to government investment. Indeed, there is an inherent conflict of interest in that the government can regulate so as to guarantee its rate of return in helium. The general cost to society of government guaranteeing its rate of return in helium is high, and perhaps provides a good reason for government to stay out of the business.

There remains the possibility that the future is much less rosy than the above assumptions have painted it. Suppose that energy scarcity and lower technological change cause per capita income to decline in the future. If so, helium will be all the more valuable since it will be one of the few important assets. This all seems doubtful, since I argued above that helium is not an essential material and will not serve to increase the costs of advanced technologies much, even if atmospheric extraction is required.

The future might be more grim than we hope or anticipate. We would do better to invest in machines, research and development, and human capital rather than store helium if that grim future is expected.

Conclusions

Helium looks like an attractive investment. I propose that we pool our pennies and set up a company to extract and store helium. But there is no reason why the government should be in the helium storage business. There is nothing essential to our national defense environment, or even our economics. Instead, the government should buy helium as needed from the private sector, just as it now buys steel or liquid oxygen.

Government should stay away from regulating helium production or prices, instead letting the free market determine price. The stored helium is a valuable asset and should be kept as a strategic reserve. The reserve could be used in time of national crisis or sold after the turn of the century for prices of $500 per Mcf or more.

DISCUSSION

ROBERT M. DRAKE, Jr.: Is there anyone in the audience who would like at this time to make an oral statement?

H. RICHARD HOWLAND, Westinghouse Research and Development Center: Any recommendations for management of the U.S. National Helium Program must be founded on the long-range outlook for the future availability of helium and uses for it. Helium is a wasting resource, in that it is a constituent of natural gas that is dissipated into the atmosphere unless extracted from the parent fuel gas.

Our reserves of helium in the shallowest, richest, oldest natural gas fields are being depleted rapidly. Though 85-100 Bcf of helium are contained in our domestic helium-rich (≥0.3 percent He) proved fuel gas reserves, substantially all these presently known reserves will have been depleted by the late 1990s.

Estimates of the amount of natural gas yet to be discovered are periodically published and revised but necessarily are uncertain. Despite continuing discoveries to date at both shallow and increasing depths, there have been no indications that additional helium-rich natural gas fields, comparable to the rapidly depleting Hugoton-Panhandle fields, exist anywhere else in the world. Some available data suggest that the helium content of the natural gas obtained from deep fields will probably be lower than that of existing shallower fields.

Estimates of helium contained in less rich (<0.3 percent He) fuel gas vary from 100 to 250 Bcf. The cost of helium extraction is inversely related to helium content. Thus

TABLE 1 Present value of helium at various discount rates.

Year	Price*	1977 Discounted Value @		
		3%	5%	10%
2000	80-90	40-46	26-29	9-10
2025	130-180	31-44	12-17	1.3-1.9

*Howland and Hulm, 1974.

extraction from these leaner streams is estimated to cost 10-20 times that encountered in existing plants processing helium-rich fuel gas. It is expected that by the turn of the century, a large, albeit uncertain, fraction of the helium contained in these less-rich fuel gas fields will also have been depleted.

Sources of helium other than fuel gas consist of proved (≥0.3 percent He) nondepleting U.S. reserves, imported helium, and air separation. Nondepleting reserves of helium are contained in shut-in (mostly low Btu) gas fields and in storage. Most of the shut-in fields are under public lands. As of late 1976, the Cliffside field contained 42.7 Bcf helium, of which 37.6 Bcf has been stored and is owned by the U.S. Government. Of the remainder, 3.8 Bcf is contained in gas native to the field, and 1.3 Bcf has been stored by private industry under storage contracts.

There are significant reserves of helium in natural gas outside the United States in the free world. The Hassai R'mel gas field in Algeria contains proved natural gas reserves of 61 trillion cubic feet (Tcf) with an average helium content of 0.17 percent, or about 100 Bcf). construction or planned indicate that natural gas production is likely to exceed 2 Tcf/yr by 1985. At this rate of gas

44

production, potential exists to extract about 3 Bcf of helium annually. (No helium extraction facilities are planned at this time.) However, the field is likely to be more than half depleted by the turn of the century, and production would decline significantly shortly thereafter. Other economically attractive world sources of helium are associated with upgrading of low-Btu gas, such as is now being done in Poland.

Eventually, with the exhaustion of world resources of natural gas, we will be forced to rely on the atmosphere as our only known continuously renewable source of helium. But the concentration of helium in the atmosphere is only 5 ppm (0.0005 percent). The cost of extracting helium from the atmosphere depends on the extent to which revenue from other valuable constituents (e.g., oxygen, nitrogen, neon, argon) would defray the cost of the air separation plant. Current demand for these major air separation products could assist the production of only a small fraction (1 to 2 percent) of the annual current demand for helium. Even then, the cost of such by-product helium would be 30-60 times the current production cost from natural gas. To increase the production of helium by air separation would tend to flood the market for oxygen, nitrogen, and other gases. This would reduce their share of the plant costs to the point where the cost of helium as the prime product would be 100 to 300 times the current cost of helium from natural gas, or $2,000-$6,000/Mcf as compared to about $20/Mcf now.

While the potential annual production of helium from U.S. fuel natural gas continually declines, the demand for helium in all likelihood will increase. Historically, U.S. government agencies and the military have been the major consumers of helium. More recently, commercial uses for helium have predominated.

Major conventional uses for helium include cryogenics, pressurization breathing mixtures, leak detection, chromatography, heat transfer and lifting gas. These uses will increase slowly in volume; the long-term trend for them follows trends in industrial production. The minimum expected cumulative demand to the end of the century is estimated to be 38-42 Bcf for conventional uses. No difficulties are foreseen in meeting demands for conventional uses to the turn of the century at not greatly increased prices.

New technologies, particularly energy technologies, that are now under development will require large quantities of helium for development and operation. The chief of these are nuclear fusion, superconducting transmission lines, and superconducting magnetic energy storage. Others include high temperature gas-cooled reactors, superconducting generators and motors, and magnetohydrodynamics. The important properties of helium for these applications are its extremely low liquefaction temperature (4.2°K or -268.9°C), and high heat transfer capacity. Unlike most conventional uses, these new technologies typically require an initial fill, with makeup required thereafter. If successful, any one of these technologies can generate energy in place of fossil fuel or increase the efficiency of energy utilization.

Estimating demand for these new technologies depends on assessment of the probability of technical success, U.S. energy requirements in the future, and rapidity and extent of commercialization. At present, none of the technologies mentioned above appears likely to have significant impact before the turn of the century. Current estimates of demand for helium from 2000-2030 A.D. cover a broad range, reflecting uncertainty. However, success in commercializing only one or two of these technologies could push cumulative demand for helium through 2030 A.D. well over 100 Bcf (ERDA 1975). A question which deserves to be answered is whether these new technologies would be priced out by helium extracted from air. Calculations show that the cost of fusion power and gas-cooled fission reactors would be increased about 5 to 10 percent if helium is assumed to be available at the price of extraction from air. The other energy technologies, with the possible exception of superconducting transmission lines, would be affected to about the same degree.

The principal uncertainties are on the supply side. Estimates of the amounts of fuel natural gas yet to be discovered have fluctuated and no doubt will be revised again. Even more uncertain is the helium reserves contained in low Btu nondepleting reserves, because there has been no profit in exploring such fields after the initial well. With these uncertainties the time when helium users must resort to helium from air varies considerably. It seems unlikely that any plausible combination of supply and demand places that time much before 2040.

Helium is presently being produced in abundance. The termination by the Secretary of the Interior in 1973 of purchase contracts under the Federal Helium Conservation Program has left production capacity in the private sector several times greater than current demand. At the time the contracts were cancelled in 1973, production capacity of all U.S. helium extraction plants was about 4.7 Bcf/yr., a rate at that point about seven times the demand. With no market, most of this production was lost by venting to the atmosphere. Depletion of feed gas sources has since permanently closed three plants and reduced potential production of the remainder to about 3.0 Bcf/yr in 1977. To reduce operating costs, the plants have ceased processing approximately one-fourth of this potential output. Another fourth is being utilized to meet current demand. Some of the remaining extracted helium is being stored in the Cliffside field under contracts with the federal government. However, a larger portion (as much as 1 Bcf annually) continues to be vented to the atmosphere. The vented production from the conservation plants wastes is in physical economic terms the lowest cost source of helium on earth. Confident estimates of (1) increasing demand, and (2) depleting feed gas streams to existing plants, indicate that demand and potential production from existing plants will cross over in 1984-1987. The cumulative excess of potential production over demand until 1984-1987 is 8 to 9 Bcf. The first two questions posed for this Forum--Should helium be conserved in the national interest? Are the present stores of helium adequate?--do not have yes-or-no answers. Rather the questions are, respectively, how much, and for what? These questions are subject to analysis in a more-or-less standard economic framework.

The rationale for the Federal Helium Conservation Program was and remains to set aside a resource now for a future time when it will be scarcer and more costly. The Federal Helium Conservation Program was set up mainly with military objectives in mind, and was not subjected to a detailed analysis of costs and benefits. The rationale for conservation now rests more upon arguments for the general welfare as expressed in economic terms. Arguments for conservation must be based on evaluations of the costs now, benefits expected in the future, and how much benefits in the future are worth today.

The management of U.S. helium reserves and stores has two relevant time-frames, with different decisions, and options. In the immediate future the question is: How much, if any, more helium does it make economic sense to conserve for future use? If it does make economic sense to store more helium, the means and tactics for doing so require attention in light of existing legislation, institutional constraints, and legal consequences.

The principal short-range government options for managing U.S. helium reserves are the status quo, government purchase of helium for storage, and inducements to the private sector to conserve helium. Under the status quo, about 1 Bcf helium is being vented to the atmosphere annually, and some plants which could extract helium are on standby status. Since the capital costs for these plants have been incurred and are sunk, the real cost of extracting helium is just their operating costs, about $4 to $6/Mcf helium. The cost of helium from new extraction plants on 0.5 percent helium content gas streams would be about $30 to

$40/Mcf, with extraction from less rich streams costing proportionately more.

Government purchase of helium from existing plants runs into a barrier not connected with social costs and benefits but relevant nonetheless. For the government to purchase helium for storage would adversely affect litigation in progress and increase substantially the risk of incurring contingent claims on the U.S. Treasury totalling about $500 million (Secretary of the Interior, 1976; Morgan, 1977). This very practical consideration cannot be ignored in any recommendation, even if the transfer of money is assumed to have no overall economic significance.

Viewed on economic merits alone, the costs of future storage at the taxpayers' expense are incurred immediately; the benefits from this storage accrue to society only when the helium stored up till now will have been exhausted. This places the beginning of the return up to the year 2000 or later, depending on how the stored helium is managed in the interim. In assessing the magnitude of the benefits to accrue one then is faced with the uncertainty in projections of supply and demand.

Even were supply and demand, and so the benefits, known with certainty in future years, the question remains as to how to value them at present. Economic resources used now for helium conservation have alternate uses which may ultimately yield greater future benefits than the same economic resources used for storage of helium. This trade-off is commonly expressed in the discount rate for future benefits. Table 1 shows the effect of different discount rates on future benefits; these benefits were assumed equal to the future price of helium, based on estimates of future supply, demand, costs of extraction, and demand elasticity. Though in each of those cases further storage appears to be economically worthwhile up to the year 2000 for helium extracted from existing plants, new construction may not be economically justified. The discount rate makes the difference. The lowest discount rate usually advocated for public projects is the real rate of return on riskless lending, historically about 3 percent. Is this rate appropriate to such a risky project as this one? The real rate of return demanded for long-lived capital investment in the private sector is about 10 percent, e.g., in the electrical utility industry for generating stations. As other authors have noted (Baumol, 1970; Hirshleifer, 1970), it seems not to make sense to remove funds yielding 10 percent from the private sector, in order to invest them in a public project yielding 3 percent.

Taking the higher interest rate as more appropriate to a long-term risky project, construction of new plants for helium storage and extraction appears marginal on its own merits. The further premium likely to be paid in the dollar transaction makes such expenditure even less attractive.

Alternatives do exist for further storage of helium, with much less contingent risk to the U.S. Treasury. One alternative is to make better use of the existing helium extraction plants by giving them a market heretofore closed, the federal agencies. Under the Helium Act of 1960, federal agencies are required to buy their helium from the Secretary of the Interior. There seems to be no reason why this market cannot be opened to the private sector. If the price for government helium is maintained, the Bureau of Mines would probably lose almost all its market, and helium extracted at the Keyes plant, about 3 Bcf further, over the remaining life of the field, would be sent to storage. Note that this is about a third of the helium that would be stored potentially from existing plants. The incremental cost to the taxpayers is the lost revenue from Bureau of Mines helium sales, about $8.7 million annually. The cost to society is unaffected, since the same plants and facilities are in place afterward as before.

Another alternative is to encourage conservation by the private sector, as is being done on a small scale presently. Encouragement can take the form of improved cash flow, less uncertainty, or both. Specific steps to increase the cash flow to companies who may wish to store helium include lowered input/output and storage rates, placing stored helium out of reach of inventory taxes, and deferring other taxes to the time helium is sold. The latter alternatives require changes in the Helium Act and perhaps other statutes as well. These actions would have virtually no effect on the U.S. Treasury, and appear to have no effect on pending litigation. The amount of uncertainty perceived by companies who might store helium concerns mainly the future disposition of the helium stored by government, as discussed further below.

The government-owned helium in the Cliffside field can be extracted rapidly and purified for sale at low unit cost. The method and timing of selling the stored helium--the management policy--will have a great effect on the market-clearing price and consumption of helium, the function of the overall market supplied by the private sector, and the benefits which accrue to the U.S. from the conservation program. There will be a management policy, if only by default. Similar effects result from leasing rights to nondepleting fields on government lands.

The Helium Act (P.L. 86-777) contains the only legislative objective for management policy; it directs the Secretary of the Interior to set the price of helium so that

revenues from sales repay expenditures for helium purchases
to the U.S. Treasury with interest by 1983, or by 1993 at
latest (Sec. 6). Other possible objectives may be, for
example to keep helium in plentiful supply at low prices, to
encourage private sector extraction of helium, to defer
depletion of the stored helium, or to maximize the present
discounted value of revenues from the stored helium. The
law of supply and demand and the structure of the
market place constraints on the objectives which can be
achieved simultaneously or at all. The more aggressively
the stored helium is sold, then the lower the market-
clearing price, the less incentive for private sector
extraction of helium, and the sooner the stored helium will
be depleted. Once the stored helium is depleted, the market
for helium reverts to free-market trends of price and
consumption.

The present management policy for release of the stored
government-owned helium may, if continued, have poor
consequences. From 1961 to the present, the Bureau of Mines
price for helium has been $35/Mcf (f.o.b. plant). This
price was set to conform to the directive in the Helium Act.
In terms of purchasing power, the real price has fallen by a
factor of more than two over this period. If maintained
this dollar price will fall below the actual cost of helium
extraction from fuel natural gas, and private enterprise
will be discouraged from adding to capacity. Once again,
the federal government would have a de facto monopoly on
helium sales, this time from storage. This would result in
depletion of the stored helium by the turn of the century,
when helium demands for new energy technology begin to
increase significantly.

The requirement in the Helium Act is obsolete, since the
assumptions on which it was predicated have been
contradicted. In principle, the government can get any
share of the market it wants by pricing the stored helium
appropriately. As to the objective, current economic
theories of economic growth with exhaustible resources
suggest that selling stored helium at a rate so as to
maximize the present discounted value of net revenues, in
the presence of competition from private industry, leads to
a socially optimum allocation of resources in the economy.
This objective has a firmer economic basis than trying to
repay the debt owed the U.S. Treasury by the Helium Act. It
begs the question of the appropriate discount rate; as
discussed before, 10 percent would be in the range.

To satisfy the objective above requires participation by
the private sector in extracting helium. The private sector
may not respond to incentives given it to extract helium.
The government-owned stored helium now constitutes a large
overhang which could conceivably be dumped on the market any

time the government chooses. Private enterprise will be
hesitant to commit funds to a project with a 10-15 year
lifetime unless prices are sufficiently high to offset the
risk of capital involved. The perceived risk can be lowered
by a clear statement of the objectives and policy that the
U.S. Government will employ in managing the stored helium.
In the absence of a clearly stated policy there could result
high prices for users, early depletion of the stored helium,
and no private sector extraction of helium.

A second problem lies in the implementation of the
policy objectives for disposition of the stored helium, and
how the implementation can adapt to changing circumstances.
A pricing method of selling the stored helium lacks
flexibility for institutional reasons. When stored helium
is being sold in competition with privately extracted helium
by a pricing method, small changes in the producers' price
may change the market share for stored helium considerably.
The purchasing power of the currency may change, and it may
be politically awkward to raise or lower prices. An auction
method, in which specified amounts of helium are sold by
competitive bid, has an advantage in that the price is set
by competition. However, there are difficulties with the
auction method too. The price will fluctuate with short-
term business cycles as well as with longer-term influences,
and the stockpile manager must distinguish between them. In
any case, preferential treatment of different classes of
customers would detract from free competition and lead to
poor allocation of resources.

References

Baumol, W.J. 1972. On the discount rate for public projects.
    In Public expenditure and policy analysis. R.H. Haveman
    and J. Margolis [Eds.]. Markham Publishing Company,
    Chicago.

Hirshleifer, J. and D.L. Shapiro. 1970. The treatment of
    risk and uncertainty. In Public expenditure and policy
    analysis. R.H. Haveman and J. Margolis [Eds.]. Markham
    Publishing Company, Chicago.

Howland, H.R. and J.K. Hulm. The Economics of helium conser-
    vation, Westinghouse Research Laboratories Report 74-8C53-
    HELIUM-R1, December, 1974 (Argonne National Laboratory
    Contract #31-109-38-2820), also in Hearings before the
    House subcommittee on energy research, development, and
    administration, May 7, 1975.

Morgan, John D. 1977. Statement for Hearings on H. Res. 91, before the subcommittee on mines and mining of the House Committee on Interior and Insular Affairs, September 20, 1977. U.S. Department of the Interior, Bureau of Mines, Washington, D.C. 35 pp.

U.S. Department of the Interior, Secretary of the Interior. 1976. Report to the Congress on matters contained in the Helium Act (Public Law 86-777, Fiscal Year 1976. October, 1976.

U.S. Energy Research and Development Administration. 1975. The energy related applications of helium (ERDA-13). E.F. Hammel [Dir.]. A report to the President and the Congress of the United States. Office of the Assistant Administrator for Conservation, U.S. Energy Research and Development Administration, Washington, D.C. 112 pp. plus appendixes.

[Discussion continued]

ARTHUR W. FRANCIS, Union Carbide: My service has included engineering research, development and design, economic analysis, market research, forecasting, planning, and product management in the areas of industrial gas and cryogenic fluid manufacture, distribution, and marketing.

The association of Union Carbide with helium formally began just 60 years ago when, on November 16, 1917, the Linde Division signed a contract with the U.S. Government to construct experimental helium plant No. 1 at Fort Worth, Texas. Good management of the helium program, or any other program of significant social consequence, should start with the regular, periodic collection of information and ideas. This should be followed by a regular, periodic disclosure to

Congress and the public of what the managers, the Bureau of Mines, see currently as the objectives of the program, and what steps they have taken and intend to take to attain those objectives. I sincerely hope that in years to come the public and the experts will be encouraged to present their information, analyses, and recommendations concerning the management of this program. I strongly urge that future sessions be organized in such a way that people have a better chance to prepare. Specifically, I recommend that when the Bureau of Mines makes its report to the nation on this subject early in 1978, a subsequent forum be convened, with sufficient time to consider the Bureau report, so that the public may criticize that report and make new presentations and offer new suggestions.

It has been said that unless we are able to learn from the mistakes of the past, we are condemned to repeat them. Therefore, a brief review of the past is called for.

The program we discuss today concerns the concepts, intentions, actions, and accomplishments of the past 20 years. In that time period, 12 plants constructed to extract helium from natural gas in North America along with four no longer existent Bureau of Mines' plants produced over 52 Bcf of helium. Of this amount, 14 billion was made available for the use of various helium-consuming processes. The remaining 38 Bcf were stored at Cliffside. The Bureau also owns or controls about 7 or 8 billion cubic feet contained in the native natural gas contained in the Keyes and Cliffside fields. In the same 20 years, the consumption of helium, which previously had been practically unique to the federal agencies, especially the Department of Defense, broadened to encompass a worldwide pattern of use, primarily by ordinary commercial enterprise. A greatly expanded distribution and marketing system developed, utilizing the most sophisticated technology to deliver helium anywhere on earth at a cost low enough to encourage continual expansion into a wide variety of new applications.

This is a very large accomplishment. Society has already obtained great value from the 14 billion cubic feet of helium consumed. In future decades, we and our descendants should obtain much greater benefits from the helium the Bureau of Mines now controls. Most commendable. We in this country have done more to serve the worldwide needs for helium than all the rest of the world put together. We are the only nation that stores helium for future use, although some other nations are beginning to study this concept. The plus side of the ledger is very substantial.

However, the negative side is also rather large. A great deal of human effort and resources was consumed to build and operate this system. The 12 plants and the storage system cost about $110 million in 1962 dollar value to construct, and about twice as much has already been spent in the operation of these plants for fuel, power, materials, and labor. A two-tier pricing system prevails, which can be maintained only by the imposition of regulations which hamper the smooth marketing of helium. This system also inhibits potential consumers, especially within government agencies, from using helium, in spite of its current abundance. The Bureau of Mines is the target of lawsuits totaling some $460 million and may be exposed to payment of about $400 million of additional sums for the helium already stored and consumed. I'm referring to the landowner and producer royalty suits.

These are impressive costs, but not necessarily excessive. However, with the benefit of 20/20 hindsight, it would appear that a substantial portion of this effort was expended needlessly. We could have provided even more helium for users at lower prices, yet hold in reserve even more than the 45 Bcf if 7 of the 12 plants had never been constructed. By following an optimum policy, at least one-fifth of the helium that was extracted could have been held in the native natural gas. The two-tier pricing system, with the associated regulatory red tape, could have been avoided. There would have been no suits against the Bureau for contract cancellation, and it is likely that the vexing question of landowner and producer royalties could also have been settled without protracted legal battles. Actual costs exceeded ideal costs by $60 million to $100 million, and included much aggravation and many legal hassles. In the same context, consider how personally wealthy each and every one of you might be now if you knew in 1957 what you have learned in the past 20 years. By way of a little humility from our own corporation, we make mistakes, and sometimes rather large ones. I'm reminded of the fact that Union Carbide was in the oil and gas producing business up until 1972, when a corporate decision was made to get out of the business -- beautiful timing.

This leads directly to the key question: How should we order our national helium program in the future? Put slightly differently: How should our natural resources, stored reserves, and production distribution marketing system be managed to augment and/or allocate and deploy those resources now and through the years ahead to achieve maximum social utility at minimum social cost in anticipation of the continually growing need for helium. First and foremost, with due consideration to specific emergencies of national security and other similar vital affairs, we should manage our helium resources economically. We support strongly the statements that have been made by the two previous speakers. By that I mean support by economic mechanisms such as price policies rather than by legal mechanisms such as prohibitions, regulations, requisitions, rationing, and the like. Perhaps the mechanism of legal fiat might have been appropriate at a time in which 90 percent of the demand came from federal agencies or their contractors. But that picture has changed. Commercial needs here and abroad now make up three-quarters of the current demand for U.S. helium supplies.

I also mean economics in the larger sense -- the sense which Dr. Lave has already discussed -- of real costs of resources, energy, and labor as well as real benefits and social utility, rather than in narrower financial terms of who pays, how much, and in what coin. For example, in the time from the start to the end of this Forum between 3 and 4 MMcf of helium will enter processing equipment that will extract the helium from natural gas incident to other operations. This helium, after being extracted, will either be vented or reinjected into the natural gas stream. All the facilities to collect, transport, and enter that helium into storage exist and are functioning. The economic cost of preserving that helium for future use is in fact vanishingly small. In contrast, the financial cost--to the operators, to the Bureau of Mines, or to some third party to conserve that helium--may be many tens of thousands of dollars for that one day's operation.

The second principle that should guide our resource management is that, at least at this point in time, only helium supply needs to be managed. Neither subsidies to stimulate consumption nor taxes, restriction, or rationing appear to be called for. Management of helium as a resource during the next two decades can be limited to meeting our growing needs from that combination of resources which most economically provides for current needs while maintaining a satisfactory reserve for the future.

In real economic terms, what are our resources? And how should we deploy them to maximize their utility? There's a storage system with some 41 billion cubic feet of helium contained in it. There are eight plants now capable of extracting helium from natural gas operating in the United States. There are two large modern extraction plants currently idle. There are gas fields containing little or no useful materials besides helium which are now shut in and which may remain for little or no cost for as long as we like. There are natural gas fields containing useful fuel now being produced and marketed that are not, but could be, equipped with helium extraction plants. Most of these fields--with about 60 percent of our known natural gas reserves--contain helium at very low concentrations well below 0.1 percent helium. About one-third of our known reserves of gas contain a modest concentration of helium, from about 0.1 percent up to about 0.3 percent. Finally, and declining rapidly, there are our gas reserves that contain helium in concentrations above 0.3 percent. Many facilities exist for air separation, for ammonia production, and for production of various other chemical products in which helium, usually in very small quantities, is concentrated sufficiently to permit, theoretically at least, some production. There exist many factories and laboratories in which helium is utilized in ways that might permit recovery and reuse. We know that vast amounts of helium exist in the air around us, albeit at very low concentration. We know that plants could be created to extract helium from the air as the sole or primary product. We know that helium is contained in natural gas fields in foreign lands, some of which could be produced in the course of their natural gas operation, and either imported into this country or used by consumers now provided with helium exported from the U.S. Finally, we strongly believe that other presently unknown deposits of natural gas will be found that contain varying amounts of helium. Most such deposits, proportionately more than for past discoveries, are expected to contain very low concentrations. Very few such new discoveries, proportionately much fewer than in the past, will perhaps contain helium at concentrations above 0.3 percent.

All in all, these constitute about a dozen known sources of helium. Each has an associated economic cost now and at various points in the future for supplying helium to meet present and future needs. The first task of good resource management is to estimate what those costs and availabilities may be. Based on these estimated costs, and some rather rough assumptions of the magnitude of future needs, a strategy or policy may then be created to deploy or hold these resources at various times in order to meet the expected needs.

This seems at first glance to be an enormous problem. In fact, it is much easier than it seems. It is so easy that I hereby recommend to this Forum and to the Bureau of Mines that an up-to-date assessment be made each year for each category. This report should roughly estimate the magnitude of the resource, the present cost of utilizing or conserving it, the changes in these expectations since the last report, the likely direction of future changes, and the intended timing, cost, and approach for applying each to meet developing needs.

I've done this myself, and I think that I won't share my own view of the matter because I really want to encourage other people to do so. I want instead to leap over the spadework, and reach an assessment of these resources and one view of how they ought to fit into the helium supply picture.

The general objective is to maximize the output of existing extraction plants, then provide incentive in the later 1980s and 1990s for end-use conservation, new construction of extraction plants operating on lean helium streams down to 0.1 percent, and also encourage at that time the production of by-product helium from air separation and chemical plants, reserving shut in and stored helium for emergencies and for use in the twenty-first century, to delay as long as possible the need for primary separation from the atmosphere. The common thread winding through all these developments is pricing strategy. The Bureau of Mines should consistently maintain its helium selling price at a level somewhat above the going price for privately produced helium. As production excesses diminish, and as new sources are required, the Bureau price should be raised to levels consistent with the required new source. For example, to return the best extraction plant for helium that ever was constructed to production to supply market needs, a price $10 to $15 a thousand higher than the commercial market pricing would be required. The Bureau should encourage this move up accordingly.

The past pricing stragey of the Bureau was not economically justified. A situation was created that encouraged the construction of seven new unneeded plants. At a time of abundant supply, new capacity was added, while at the same time demand was inhibited by excessively high prices. The danger in future years is that through inertia, and lack of clear insight into the power of the price mechanism and the goal that ought to be pursued, the Bureau may retain the $35 price until inflation and inherently rising costs of private producers combine to make what was once much too high a price much too low. At that point, as helium actually shifts from abundance towards scarcity, the then low price would forestall new construction of extraction plants while encouraging wasteful use. The stored helium would quickly be consumed, and halfway into the next 60 years a horrendous supply crunch would develop just as our grandchildren desired to make large-scale use of the wonderful new technologies we so carefully provided for them.

If you think that no intelligent public servants could possibly pursue such an idiotic course, recall that the Bureau has already presided over the first part of this economic nightmare, and the Federal Power Commission enacted a travesty similar to the second part in controlling natural gas fuel prices through the past 10 years. To avoid this, we recommend that the Bureau start now to adjust its helium prices each year to reflect the change in monetary value, and that in addition, the Bureau management add further increases as continued study indicates so that full use is made of all desirable depleting helium sources before stored helium is withdrawn for use. In addition, we recommend that the various government agencies having control of natural gas production from federal lands restrain the development of shut-in helium-bearing gas reserves until all sources of 0.1 percent helium content fuel gas streams are effectively utilized.

Another important step in maximizing the social utility of helium at the lowest cost would be to end the requirement that federal agencies purchase their helium needs from the Bureau. There is no need to reverse this restriction as has been suggested in the legislation proposed to the Senate by requiring the agencies to buy privately produced helium. As long as the Bureau maintains a pricing incentive to encourage private production, most of the government agency users will gravitate toward the lower price from private industry. By eliminating the reserved federal market, the need for helium regulations would be eliminated, and they should be withdrawn. This would also reduce the amount of litigation. Furthermore, the Bureau would then have no need to proceed with the planned entry into liquid helium production in competition with adequate existing privately owned facilities.

Insofar as the Bureau does perform purification, liquefaction, transportation, packaging, and marketing services for helium users in direct competition to existing private industry, the Bureau prices and terms for these services should reflect the cost and profit required to maintain and expand such services so as not to compete unfairly with the private taxpaying entrepreneur. The Bureau should support the proposed legislation to permit expensing all costs associated with the private placement of helium into long-term storage in order to remove this potentially inhibiting financial disincentive to private storage.

Finally, and by no means least in importance, the Bureau management and legal staff should expend every effort to assist the federal district and circuit courts to reach an economically viable conclusion to the landowner and producer royalty cases. The greatest weakness of the Bureau's policies has been to depend on legal steps to accomplish what good economic planning implemented through sound pricing policy could have accomplished more easily. Future Bureau policy should eliminate regulations and legal restrictions, and should pursue economically sound goals through appropriate pricing. Those goals should include the maximum use of existing plant capacity to extract helium from depleting helium-rich fuel gas deposits. As these

s, the financial cost--I'm not talking about economic
--would be horrendous. Prudent management of
holders' interests precludes taking any kind of gamble
the courts can resolve this issue. One of the crucial
ys the government can do is to try, insofar as it is
ly possible, to expedite a reasonable settlement of the
owner and producer royalty case.

I believe that Cities Service has somewhat the same kind
roblem. Although, they're involved in a different case,
burden may not be quite so large even though they face
same kind of problem. I'd like to go back in history
put my finger on the timing of what I think is the most
rtunate circumstance. When, finally, the decision was
hed by the Secretary of the Interior, after the
ronmental statement had been issued, to terminate
age, the various companies involved had already
icated to the government their desire to continue to
r helium physically into the storage system. Reserve
question of who owned that helium and who would have to
for it and at what price for the completion of the
igation. Upon determination of the cases the appropriate
ments would be made and ownership of the helium would be
ablished by the decree of the courts. Unfortunately, in
estimation, the government chose not to do this and, in
t, the judge involved in the cases could not, on his own,
e action to order this. The question of how this helium
be stored is a difficult one. This may take some subtle
euvering--maybe some private agreements amongst
ganizations--but it seems to me that first and foremost is
e issue of royalty payments. I'd like to emphasize that
cause it is one area that is often overlooked.

The question about the value of helium in native gas
reams before extraction is something nobody has tackled
th any degree of sophistication. There is no value
trinsically to helium. Helium is absolutely worthless
til a user establishes the value by finding a way to put
to use and creates a demand. The value is determined by
emand and demand has to be considered in a time frame.
en you have a small current demand and a large current
upply, it strikes me as questionable if we find high
conomic value for the element in the raw material state.
'd like to elaborate on this and identify some remarks that
made previously to illustrate how they are pertinent.

I made the claim that 7 of the 12 plants were
onstructed uselessly. One of those was the Keyes plant of
he Bureau of Mines--the first of the 12 to be constructed.
t was ordered and constructed in the same general time
rame as the start of the conservation program. The field
from which Keyes draws its gas is rich in terms of helium
content. It's a better than 2 percent field, and one can

decline, prices should be encouraged to rise enough to
permit extraction of by-product helium and construction of
new plant capacity to process leaner helium content fuel
gas. Naturally conserved helium deposits and the bulk of
the stored helium should be reserved until market demand
exceeds these sources. Every effort should be made to
forestall as long as possible the need to extract helium as
a primary product from the atmosphere. Excessively high
prices during the past period of abundant supply restricted
use, brought regulation and legal battles. Unreasonably low
prices during the coming age of scarcity will restrict
production, encourage waste, and will consume our precious
resources before our descendants have the proper opportunity
to enjoy the fruits of our labors. If we let that happen,
the judgment of history will be harsh indeed. We can do
much better, and we should determine to do so.

FRANKLIN A. LONG, Cornell University: As a member of the
Committee I've been impressed and encouraged by the
statements so far. But I am now at the point where I would
like to hear some comments from the viewpoint of the Bureau
of Mines.

To the economists who have talked to us I have a
particular question: Other than the legislation, is there
anything negative about firm and indefinite withdrawal of
Bureau supplies of helium from the market? If Bureau stores
of helium were treated by legislation as a sealed-in reserve
for some indefinite time, would that in any way be
troublesome?

LAVE: I don't see any reason offhand why that would be bad.
If the government reserves were sealed in, then the whole
current market would have to come from private producers,
which would certainly encourage some of these closed-down
plants to open up. One can't ignore that 41 Bcf. There has
to be some policy for when it's going to be sold off. It
could be reserved until after the end of the century, but
Mr. Howland's statements make a lot of sense. We should
have a policy that maximizes the present discounted value of
all that helium.

I must say that I'm somewhat suspicious of Mr. Francis'
proposal since that really requires that the people at the
Bureau be very farsighted indeed. Although I'm willing to
believe that the people at the Bureau are intellectual
superpeople, I'm not so sure that they always have a free
hand in doing these things. So there is a lot to be said
for cutting down the amount of discretion to be exercised
and for not having them make these forecasts about what's
going to happen to technology and supply and demand over the
next year, but to try and state in advance a well-set-out
policy so that the private producers know what to expect in

future years. I am really attracted by Mr. Howland's proposal that at various intervals the government resource will be auctioned off beginning, say, in 1985 or 1990.

HOWLAND: I agree with Professor Lave. If we shut in the Cliffside field, and in the short term send all production from government-owned plants into storage, that's great up until perhaps 1987. But the essential question is: Having stored the helium, when do we begin to get a return from conservation? We need objectives and methods.

NICHOLSON: I would like to go with Dr. Long and to urge that the other side of this argument be presented. Why did we set the price as we did in the Bureau of Mines? It seems that the answer has to be clear when we consider alternatives we might pursue. I don't think from what we've heard so far that we have any idea or any defense of the action that's been taken by the Bureau of Mines over these many years.

HOWE: I'd like to ask Dr. Howland what his position is with regard to resources currently being vented? He indicated that continuing storage has no real economic role to play at present. We do have this issue of currently operational plants that are venting a resource. Marginal cost recovery is essentially zero, or very close to it. What would you recommend with regard to the allocation of this resource or the program for its use if we could ignore the potential legal implications or the legal issues that might be wrapped up in the Bureau's current attitude toward this resource?

HOWLAND: I have tried to take some possible future market clearing prices for helium and discount them back to the present at various discount rates to illustrate the effect these rates might have. Taking real discount rates of 3, 5, and even 10 percent, it seems to make economic sense to set aside the helium currently being extracted by the existing plants. If one talks about new construction of additional plants right now for the purpose of storage alone, that's another question. I would just as soon not talk about new plants in regard to your question. However, we cannot ignore the legal legacy, and I don't think that any proposal to store more helium at the taxpayers' expense is going to fly with Congress. Too many people know what the program has done so far in terms of fiscal objectives set for it in the first place. Next, so far as the legal legacy itself is concerned, one of the defenses employed by the Justice Department in fighting breach of contract suits is that we don't need to store the helium. If the Congress were to pass legislation authorizing or demanding further purchase of helium at taxpayers' expense, it would substantially increase the possibility that companies suing the government for breach of contract would win. These contingent

liabilities are something like $400 mi
economists we can argue that just to t
place to the other really doesn't have
physical allocation of resources in the
gripe about transferring money from tax
stockholders of companies doing the sui
legislators the contingent liability wo
frightening. The administration now is
dimes and nickels and probably can't fir
budget to authorize any new program. I
the fundamental economic merits, further
definitely justified. Done at the taxpa
run into a can of worms. That's why my
in the direction of trying to encourage
store helium at their own risk and expens
very little effect upon the litigation in
would cost the taxpayers virtually nothin

JOHN K. HULM, Westinghouse Research and De
As I understand it, of the four plants tha
with the Bureau of Mines to supply helium
1961, one plant is currently recovering he
it in the Cliffside field but the governme
for it. The helium still belongs to the c
to a report from the Bureau of Mines. In t
the natural gases ran out, and the plant is
other plants are still running, but the hel
vented. In 1975 the Bureau did make arrang
the gas on behalf of the private owners of
I'm puzzled as to why only one of the operat
advantage of that arrangement. The other tw
venting. We're losing a billion cubic feet
helium that could be stored--as each speaker
no marginal economic costs. There seem to b
other obstacles preventing this storage. I
you could enlighten us on the hangup here. W
crucial question? Why don't they open the va
the helium go into storage?

FRANCIS: People from both Phillips and Cities
answer that question better than I can. Howev
be embarrassing for them to do so. As a compl
I'm going to make all sorts of assumptions. I
most critical problem of all is the nature of
litigations that we're involved with. Litigat:
respect to landowner and producer royalty payme
a problem to Phillips in respect to helium they
produced and sold to the U.S. Government. The
decisions rendered by the courts would indicate
are liable to pay between $12 and $17 a thousan
landowners and producers. If that same type of
payment were, by law and the interpretation of t
also placed against helium privately stored by t

show that it would have been substantially cheaper for the Bureau of Mines to purchase the whole field and hold it out of production than to produce from it, pay money to extract helium, and stick the helium back in the ground. This anomaly occurred because of a lack of appreciation of the intrinsic value of helium. That is, what is the value of helium in natural resources. If one is going to buy helium at $10 to $12 a thousand and put it into ground storage for future use, you can show that the value of the helium in the Keyes field, or the Greenwood field, or Pinta Dome that Kerr-McGee developed is of greater value than the fuel content of that gas because fuel was selling at that time for $0.10 a thousand at the wellhead and the helium content must have been worth at least $0.20 a thousand in the same gas.

One of the things that should develop in the coming years--I don't say in the coming months because it takes a little longer than that to get the facts together and marshal the ideas--is an evaluation of helium in native gas streams. We have these plants, some of which cost a great deal. The national plant, for example, cost $25 million, and it would take $50 million to duplicate. We have streams of gas flowing to market that are not utilized. In the future we need to maximize the value of helium to society by redirecting streams of higher helium-content gas into existing plants to be processed while swapping out low helium-content gas so that fuel markets remain supplied. This is the function of royalty payments. This is where the value of helium in a flowing gas stream becomes important. These things should be analyzed; they should be considered. As soon as possible, we should get a determination from the courts giving some rational basis for royalties.

HULM: I find your remarks very attractive because your suggestion would make some sense out of the mess that exists in the royalty area. Are you suggesting--when you say this should be done--that the Bureau of Mines should make an analysis and set some kind of value on royalty payments? Is that what you have in mind? If not the Bureau of Mines, who's going to do this?

FRANCIS: I'm going to beg that issue. But if the Bureau or any other government agency is going to tackle the question of resource management in regard to helium, then this is one of the questions that they must tackle. I'm not going to make any more recommendation than that.

LeROY CULBERTSON, Phillips Petroleum Company: As Vice President of Natural Resources Group Planning and Budgeting of Phillips, I wish to address the question asked about why more of the companies that were in the helium conservation program are not jumping at the "opportunity" to store helium in Cliffside storage.

When the Congress of the United States passed the Helium Act amendments of 1960, there was a need to conserve helium and the Act was a reasonable solution to the problem. The natural gas containing helium was being produced and consumed at a rapid rate and the contained helium was being lost into the earth's atmosphere.

Natural gas from the large Hugoton and Panhandle fields of southwestern Kansas and panhandles of Oklahoma and Texas generally has a helium content in excess of 0.3 percent which has been considered by the Bureau of Mines to be the lower limit of economically extractable helium. Until the helium conservation program was terminated in 1973, about one-half of the gas from these fields was processed for the extraction of helium. Today, due to the strong demand for natural gas, these fields continue to produce at maximum allowable rates. Since the government conservation program is not in effect, much of the helium which could be relatively easily extracted and saved is being lost into the atmosphere.

The United States is importing about 40 percent of its crude oil now and the percentage is growing. Energy sources are changing and new technologies are developing. In the future, helium may be essential as a source of temperatures low enough to permit superconductivity in electric power generation and use. Nuclear fusion, gas cooled nuclear fission, superconducting power storage, and magnetohydrodynamics all use helium. It is my judgment that the helium conservation program was a wise one and its cancellation was a mistake.

At the end of 1973 the Panhandle and Hugoton area gas fields contained about 97 Bcf of helium contained in natural gas. At the end of 1976, these fields contained only 81 Bcf of helium--a depletion of 16 Bcf in three years. Under the conservation program, about 50 percent of this helium would have been conserved. As it is, relatively little of this helium has been saved. Most if not all of this gas will have been depleted before the end of this century.

When the Helium Act amendments of 1960 were passed, the demand for helium was supplied almost entirely by government activities. As a consequence of the Helium Act amendments, private industry built facilities which recovered 3.0 to 3.5 Bcf per year of helium that was sold to the government. This was at a time when total U.S. demand was only 0.6 to 0.9 Bcf per year. Today government holds some 40 Bcf of helium in storage; private industry and government currently can produce at least 2.4 Bcf per year; while the market needs less than 0.9 Bcf per year.

All the companies in a position to extract helium are confronted with a dilemma. Obviously, the present helium markets can be supplied for many years into the future with only a fraction of the helium which could be extracted and saved. The helium now in government storage ensures an adequate supply for many years. Therefore, most of any additional production of helium would have to go into long-term storage to await future anticipated increases in helium demand.

It is my opinion that currently there are several impediments to the storage of helium by private industry:

1.   The government can control the future price of helium by virtue of its large supply. Long-term projection of future prices represent a major risk to private industry.

2.   Costs of extracting and transporting helium stored by private industry cannot be expensed for income tax purposes until the helium is sold.

3.   Ad valorem taxes levied by states on privately owned stored helium could significantly affect the economics of storing helium.

4.   Current litigation in state and federal courts may result in helium extractors being required to make additional payments for helium to natural gas producers. One U.S. district court held that Phillips must pay a natural gas producer $12 to $17 per 1,000 cubic feet for helium extracted and sold to the United States. This decision has been remanded for further hearing on the valuation issue. The uncertainty of the liability, if any, for additional payments to gas producers and the amount of such additional payments, if required, for helium extracted in the future, is a serious impediment to its extraction and storage.

5.   The government and Phillips have agreed on terms for the storage of privately-owned helium in the government Cliffside storage field near Amarillo, Texas. The contract covers a term of 20 years, with the option to extend for 5 more years. This is not long enough for a private investor engaged in long-term storage since he must assume the risk that some of the helium could remain in storage much longer than 25 years before being needed to satisfy then existing markets.

By December 1, Phillips expects to be delivering helium to the government for storage under the terms of the storage contract at the rate of about 0.3 Bcf per year from our Sherman plant. Our Dumas helium plant is not extracting helium due to lack of an adequate volume of feedstock gas. The extent of our continued use of the Cliffside storage contract will be influenced by the manner in which the above impediments are resolved.

The Congress should act again to maximize conservation of helium, but in doing so, it should recognize as it did in enacting the Helium Act amendments of 1960 that the recovery and long-term storage of helium can best be accomplished through the joint efforts of government and private industry. The Congress should act swiftly to save and should encourage private industry to save as much as feasible of the large volumes of readily recoverable helium being dissipated without serving any useful purpose.

After this helium is gone, the nation will have to get new supplies from more costly and uncertain sources such as low-helium-content gas which, as a last resort, would be the atmosphere. According to House Resolution 91, ERDA has concluded that the economic cost of extraction of helium from the atmosphere is $3,000 to $6,000 per Mcf, and that the energy cost of producing 1 Bcf per year of helium from the atmosphere would require at least 50 percent of the projected annual output of the Alaskan oil pipeline.

There are several steps that I believe can be taken and will result in increased conservation of helium. Senate Bill S.2109 recently introduced by Senator Pearson will accomplish most of these steps if approved by Congress as it was introduced.

1. Removal of the five impediments listed earlier would encourage the long-term storage of helium by private industry. Most of these impediments can be removed by legislation.

2. To conserve the present stores of government-owned helium now in Cliffside, and to allow conservation of all of the helium produced from government plants in the future, federal agencies should obtain their major helium requirements from non-federal sources. The current law requires federal agency major helium requirements to be supplied from the Bureau of Mines. The government should conserve its helium until private industry can no longer reasonably supply current demands from depleting natural gas. This action would effectively conserve about 0.15 to 0.2 Bcf per year of helium. This change could be accomplished within a reasonable amount of time.

3. Another step that could be taken to further conserve helium would be for the government to reinstitute

the purchase of helium for storage. Plants which currently have excess producing capacity obviously would be immediate sources for such helium purchases. In addition, there are some helium-bearing natural gas streams flowing to markets from the Hugoton and Panhandle fields that have never been processed for helium extraction. Some of these streams contain in excess of 0.3 percent, and the helium extracted could be relatively inexpensive as compared to alternate future sources. Consideration should be given to the extraction and storage of some of this helium.

LONG: Is that 0.3 Bcf per year you mentioned the capacity of the Sherman plant, or is it what you've decided is an appropriate ante in this particular business?

CULBERTSON: I'm advised that is our capacity.

V. KERRY SMITH, Resources for the Future: It seems to me that the gentleman from Phillips has indicated a number of disincentives to storage. Professor Lave on the other hand told us that there are substantial reasons why private firms might be willing to store helium. Dr. Howland gave us more or less the same message: there ought to be incentives for helium storage. Third parties, including the government, are intervening to affect future prospects, understandings, and expectations for and of the demand and supply situation for helium, particularly the prices and terms for long-term storage. Are there impediments other than those identified by Mr. Culbertson from Phillips?

FRANCIS: I think that Mr. Culbertson's statement did include all the impediments that we can see at the moment. I would like to make it clear that we're making progress. The government was formerly criticized for not having a proper policy with respect to storage. They responded to that criticism. Maybe they took too long to respond, but this is not pertinent. They did reach a new policy with respect to storage contracts. They consulted people who had shown an interest in getting contracts, and they wrote some contracts that were attractive. Subsequently, several private parties acquired helium and utilized those contracts. What we're hoping for as a positive output from this Forum is continued progress by the government. Some remarks both yesterday and today may seem to be rather critical of the government. We should be very careful to note that this criticism is and has been constructive and that we expect progress. I think that there is a good expectation that these impediments can be removed and that the result of their removal will be significant private storage.

HULM: Is there someone here from Cities? It is very interesting to learn that the gas from the Sherman field will hopefully be conserved fairly soon if Phillips reaches agreement with the Bureau. I wonder if any negotiations are in progress in the Cities Service's Jayhawk field where 0.7 Bcf of helium could be produced annually.

GEORGE C. VAUGHAN, Cities Service Company: In respect to the helium program, our plant at Jayhawk is one of the conservation plants. Venting from the plant is something over half a Bcf a year, and we would very much like to see an atmosphere in which this could be conserved. The impediments to the conservation and storage of helium have been pointed out. The long-term nature of the conservation program is not something that a private corporation can handle.

The helium program has been a wonderful thing. If it had never been undertaken, we wouldn't have this 32 Bcf at Cliffside. It is regrettable that it was stopped. I do think it is a program that must be carried out with government leadership until we can resolve the long term nature of return on investment. Until we can resolve the landowners' claims it will be very difficult for us to take any private action to put the helium in storage.

I think it behooves this Committee to think in terms of the way we see our energy position today as compared to the 1960s and in terms of the usefulness of helium and its contribution in efficient use of energy. Helium use may be even more critical than it ever has been in the past as we turn to a world of alternate fuels. I wonder what has been the real economic cost of this storage program. If there have not been great profits to the companies with the conservation plants, one-half of whatever profit there was has gone to the federal government. This would be true under the continuation of any program. In private helium storage programs, we must seek a way to remove these impediments.

LAVERICK: As must be obvious by now, there are a large number of men of good will trying to resolve this very complex helium controversy. I'd just like to make one or two remarks now in the context of the present discussion.

I agree with Lester Lave's philosophical discussion and expressed similar sentiments in my long 1974 preliminary NSF report. However as a first instance we are dealing principally with a specific problem--the conservation plan. In a climate where the government has cancelled the storage contracts for helium-rich gases of the Hugoton-Panhandle fields and the controversy has raged for eight years, we never even thought, as Richard Howland mentioned, of raising the question of extending storage to gases where the helium content was less than 0.3 percent or in gas fields with less gas than those defined by the contract.

Clarence Kipps said yesterday that the government is going to lose its cases with respect to the contracts. I am sure that the government people might question that. If I were on the government side I might question it. The fact remains that in the case that was tried in the Court of Claims by the Full Court, the amount of money involved was reduced by half and the matter was sent back to the trial judge to see whether it could be reduced any more. In other words, the argument was about money. It wasn't about whether or not Northern Helex was right. The court seems to have granted that. There is the danger that the helium that is being thrown away will be paid for again by the government, by the people. It will be paid for twice.

What Lester Lave said that might make good sense from the economic point of view was that maybe the government should not have entered into the helium storage contracts in the first place. Maybe he is right. Maybe we should not have killed 50 million people in World War II. From the economic point of view it didn't make any sense. The reason for storing the helium at the time was because we had the liquid-fueled ICBMs and we needed the helium. So there were reasons other than economics for storing that helium. The fact is that the helium is being stored; the plants are there; they are operational; they have been paid for; the helium is going up into the atmosphere with the plants running when all it takes is somebody to turn a valve to store it underground. They do need some money to operate the storage compressors. Both the ERDA and the NSF reports concluded that this made good sense.

Lester Lave pointed out that investments in research and development have paid off best in the interval he discussed. It should be obvious that companies with no research and development go down and the same applies to nations. This is a very advanced society; we live in dangerous times. MAD (Mutually Assured Destruction) is part of the policy of the United States and the Soviet Union. I think that, as a whole, United States society is going to have some trouble getting through the next 50 or 60 years as will our whole advanced civilization.

The comment about future generations bothers me. I wonder what a future generation is. If I were a teenager today, or say 20 years old, and if I got my biblical term of three score years and ten, I would have another 50 years to go. That's almost up to 2025. If I'm a little five-year-old at school, maybe I have got 65 years to go. That is to 2040. Is that future generations or is it today's generation? They are alive; they are here.

On the question of new technologies, yes, fusion is pie in the sky. However, if it wasn't for some dreamers that tried to turn pie-in-the-sky ideas into reality, you wouldn't have such things as television, radar, motor cars, and then maybe you would need those horses collars that Lester Lave talked about. The pie-in-the-sky people are developing superconducting magnetic energy storage, underground superconducting power transmission, and fusion power with confining fields using superconducting magnets.

You've heard why the gas companies find it difficult to store. You've heard about the tremendous debt that is owed by the government to the government for the government and that is 41 Bcf of helium in hock down in the ground which at today's prices is worth about $820 million if it could be sold. The debt is $430 million.

HOWLAND: In the twenties there was a scare when the supply of whale oil began to go down and people were worried because they didn't know where whale oil was going to come from. That problem seems to have cured itself.

M. KING HUBBERT, consultant: I just heard the remark about how wrong we were in our oil and gas estimates in the past. I'd like to remind you that we passed the peak of oil production in the United States in 1970 and gas in 1973. It is most unlikely that these two peaks will ever be reached again.

SMITH: It seems to me that the individuals involved aren't disagreeing about conservation, but about who is conserving? The prospective demands that you've indicated exist are sufficient to ensure that there are private investors who, if these obstacles are removed, will be willing to undertake the risks. Will we pay for conservation with taxpayers' money and let government assume the risk? The issue is one of whether there are social benefits associated with the government's assumption of these risks. It seems to me that benefits associated with the conservation of helium are in large part capturable by private interests, if they're associated with essential technologies to produce energy materials.

HOWLAND: Professor Lave and I guessed that private enterprise would find it attractive to store helium now for use later. We do not know the individual finances of the companies well enough to make it a sure thing. The history to date is somewhat encouraging. There are a few storage contracts. Phillips Petroleum has recently made up its mind to start storing some helium -- even in the face of all the uncertainties as to both litigation and future government policy. It would seem offhand that the private sector does have some interest in helium. What we're trying to do is to establish what the impediments or disincentives are and remove them.

LAVE: I'd like to thank Mr. Laverick. I imagine that parts of what I said that were set up for barbs were the parts that I hadn't cribbed from his report. First of all, I'd like to say something about future generations and the ability of the private market to invest in things that may be uncertain and far in the future. I'd like to ask perhaps the rhetorical question of how often any of you have drunk Scotch that was distilled in this year or how many of you have ever had a bottle of Lafite-Rothschild that was bottled within six months of when you drank it. If we have sufficient funds, we can buy brandy that's more than a century old. If you would like to have a 10-year-old good Bordeaux or Burgundy you need not put it down yourself; you may go to any one of hundreds of establishments within a few miles and buy it. If you have any doubts about the ability of the private market to make investments with large amounts of uncertainty over long periods of time, go out and try and acquire some oil shale land or land with coal under it and see whether that sells simply for the value of the surface rights. Even though there is no commercial technology for oil shale and although we have, by anybody's estimation, a lot of coal, you'll find that the mineral rights there sell for considerable sums of money. Somebody seems to be willing to hold assets in this very long-term way even with all of the government-engendered uncertainty as to what those assets will be worth when it's time to sell them. The oil shale might be worth essentially nothing because the government could either commandeer it or could regulate the price so that it was worth nothing.

I have some advice for the Committee. Put first things first. What are helium extraction and storage worth to society? Insofar as that question gets settled in the affirmative, by saying that it is worth a great deal to society, then the Committee can take a look at the next question: How it is that you can get that valve turned so that helium goes underground right now? I really think those questions have to be separated. One is relatively easy to answer. The second one is almost impossible to answer. It seems to me that I have heard what Mr. Culbertson said before. I read it in a novel called Catch-22. I take it that none of us on the Committee is going to try and get through that horrendous set of regulatory and legal problems that talk in detail about how you get the switch turned on. There might be some suggestions, but that's really something that a group of well-paid lawyers -- I assure you they will be -- can get at. If the Forum can clarify or identify some of the problems, then there is a great deal to be said for it.

Another point has to do with research and development and where it is going. So far we have heard about some of the R&D that might provide us with marvelous uses for helium in the future. I agree that we ought not to be shortsighted and say that any use of helium that is prospective for the twenty-first century is simply beyond our vision and should be ignored. But R&D can go on in methods of re-extracting the helium that is used and putting it back for further use again -- recycling the helium. R&D is undoubtedly going to find cheaper ways of extracting helium either from natural gas or from the atmosphere. As I understand it, there is no theoretical reason why one has to expend all of the energy required to liquefy the atmosphere in order to get the helium out of it. If that is so, then might it not be true that somebody is going to come up with a jim-dandy way of getting helium out of the atmosphere that might require considerably less energy than liquefying it?

As I have said, I would personally be happy to put my own funds in helium storage. Helium is a good investment. Given all of the wild uncertainties with respect to future government regulation and so on, it still would earn a handsome rate of return.

Should we be conserving helium as a society? If the answer to that is yes, then perhaps one should ask why the current situation is so muddled. I must say from my wanderings around the government looking at policies that that question is so complicated that I doubt that any committee of the National Research Council is going to shed a lot of light on why the situation is so muddled, much less what one can do to get us out of the mess. I think that the private market is perfectly capable of handling all this.

I do want to react to Mr. Laverick's comment saying that there was a role for the government in this early on. It has to do with what we economists call the infant industry argument. That argument is that we are all not infinitely farsighted, and so what has to occur in many industries is enough initial technological and financial interest to get them started. Sometimes that's provided by a very farsighted investor who has considerable money; other times it is provided by the government. In the case of helium, I would guess that the government took a major role in making helium available so that people could think about serious commercial uses for helium. The infant industry of helium is now at least a respectable strong teenager and quite able to go on on its own. Indeed, I would think that further parental care in this case is probably going to stop growth, not speed it onward.

So I again come up with this overall recommendation that the whole situation could be managed very well if the government could somehow or other be gotten out of it. One of the few things I disagreed with in Mr. Howland's statement had to do with whether government actions could render nil the recommendation that private storage looked attractive and he said that it looked like it could. Even though there are 40 Bcf of helium or so in storage at the moment, it seems to me that any reasonable projection would have that running out by about the end of the century, in which case the price is likely to go up and helium in private storage would be called out at attractive prices. It would be much more attractive for private storage if one could depend on selling it before the end of the century, but that's all right. Independent of the government's actions, all of that government helium is going to be sold sometime between now and about 2020. I would think that in any private investment calculation it makes sense to buy and store helium now quite independently of what the government is going to do, as long as the dead hand of price setting doesn't get in there to keep the price of helium down at low levels to "encourage use" and thereby stifle the whole market.

HOWLAND: There are varying degrees of attractiveness for private storage, one of which is how soon a company thinks it can get its money back. The policies that the government follows in the future for pricing or selling off the stored helium will make a difference in what the private sector does. It certainly will have a great short-term influence. Art Francis has suggested on the one hand, that the government always prices itself out of the market essentially. On the other hand, we have a pretty good idea that if government keeps the price of helium at $35/Mcf, it will have all the market, by about 1990.

LAVE: I have a sort of a radical proposal that the government put all of its 40 Bcf on the market right now. I'll go in for a share whether they auction it off at $440 million or $800 million, that is either at cost-in -- to use the accountant's notion -- or at current market prices. I'll buy it at either one of those two prices, or I'll buy my share of it. I would argue that helium investments are going to have a handsome rate of return. This may have the salutary effect of getting the government out of the helium business once and for all, and it might make the market look very nice indeed.

I certainly don't disagree with the statement that the government could make policies which would encourage more private storage now such as deciding to hold off that 40 MMcf for some period of time in the future. But I would say that, if we had a careful analysis of the situation by the private companies, they ought to conclude that whatever the government action, it is attractive for them to store helium

and not simply helium at zero marginal cost. I think it is
only legal difficulties that are in the way of private
storage now. I'm not quite so sanguine about how soon those
can be resolved, but perhaps there is some other
institutional mechanism that one could come up with for
storing helium in the meantime.

LONG: When one has an expert like King Hubbert here, one
ought to take advantage of it. My question to him is:
Have we as a nation and as a world system really analyzed
the helium supply problem adequately? Go back to the
earliest days of uranium use when the alleged supply of
uranium was very small, because as someone pointed out, no
one had been in the slightest interested in looking for it.
Do we have any sense that where we happen to stand right now
with helium represents the best that we're going to be able
to do in the way of finding helium reserves?

HUBBERT: Our knowledge right now about helium, especially
in the United States, is that the accessible helium occurs
in natural gas in the various percentages that have been
discussed this morning. We know what these percentages are
in various gas fields, and we also know approximately how
much natural gas we have. The evidence that we have now for
the United States adds up to an ultimate amount of natural
gas for the lower 48 states of probably about 1,050 to 1,100
Tcf including past production. If we take out past
production -- and I don't remember what it is right at the
minute -- 400 or 500 Tcf, it leaves us somewhat less than
half what we had when we started. In the lower 48 states
we've already passed our gas peak in 1973 and all the
evidence we have is that our natural gas will be almost
exhausted by the year 2000. To put it another way, the
middle 80 percent of our ultimate gas production has covered
a span of about 65 or 70 years, the midpoint of which is
about 1973. Since helium is tied up with natural gas we
have correspondingly reliable estimates on helium. If we
keep on with the types of activities that we're in now, most
of that helium will be in the atmosphere before we get
around to doing anything effective.

A.D. McINTURFF, Brookhaven National Laboratory: My major
concern and the real reason I am at this meeting is that we
now have an energy crisis or a possible shortage. Certainly
we can see limits to our energy supply, and there are in the
physical sciences certain technologies that point toward
high degrees of efficiency in energy use or at least toward
greater efficiency in its use for things that now can be
done. Some of these that seem to be developing are very
high energy density devices, and devices that seem to be
highly efficient are built on a technology of helium, not
necessarily liquid helium, but near-liquid helium
temperatures. Of course the demand for and the application

period for these high-energy density and efficiency devices
-- as everybody has reiterated many times this morning --
isn't until the year 2000. However, there was a thing in
physics that lasted for a long time called Carnot cycle
which most helium refrigerators are usually governed by. It
is theoretically difficult to beat that cycle. I think we
will not improve present technology to any great degree --
maybe one order of magnitude -- so maybe we can get down to
the 8,000 if we are extremely clever or much more clever
than our predecessors.

B.W. BIRMINGHAM, National Bureau of Standards Laboratories,
Boulder, Colorado: The motivation, as I originally
understood it, for the helium conservation program was the
increasing use of helium in the late 1950s, as I think Mr.
Laverick said, for intercontinental ballistic missile
projects. At that time some of the new technologies that
will be talked about later today were not in the picture, or
at least we didn't know how to harness them. I wonder if
the economic equation for conservation hasn't changed.
Maybe the mistake was in not having this Forum in 1970 or
1971 before the Department of the Interior elected to cancel
the conservation contracts. But my recollection is that it
was indeed a federally inspired activity, particularly in
connection with the aerospace program, that there was not
the private incentive which conceivably could exist today.
People have mentioned superconductivity and that the
discovery of superconductivity and the Wright brothers
activities occurred at about the same time. But unlike
flight, superconductivity has not come of age. I think the
breakthrough only occurred about the time the conservation
plants spoken of here were put into operation in the early
1960s.

LAVE: I quote a line from Stephen Foster, "I bet my money
on the bob-tailed nag, somebody bet on the bay." There's no
certainty in looking at the future. The only way we resolve
questions of the future is by trying to consult our crystal
balls. You have two very different approaches. One is that
of the socialist countries where if you are going to make a
bet on the future it has to be duly approved by everybody
all along the line. If we take a look at the rates of
research and development there and the implementation of new
ideas, they don't seem to be startlingly rapid. Whereas, in
the United States there are literally thousands of damn
fools who are ready to put up their hard earned money to
test out some idea that might conceivably make them rich.
And as a result of that most of those damn fools go back and
lick their wounds later on, having lost whatever they had.
But every once in a while we come up with these magnificent
successes that we have around us. From a societal
perspective you can ask whether that's good or bad.
Although the uncertainty is vast about whether the

superconducting technologies or fusion or anything else will ever be of commercial interest, I was one of the damn fools who was willing to put my money on the line. I think that if what we ask for is a full governmental debate to try and convince the Congress beyond reasonable doubt that helium storage is worthwhile then we are likely to be engaged in that process at the end of the twenty-first century, when by anybody's estimation the helium is all gone. That may be one of the difficulties we have already experienced in the helium conservation program.

You know it turns out consistently that the economists have more faith in technological change than the technicians do -- it's probably safer to say the academic technicians -- since again you have all these private damn fools who work in their laboratories and put up their money and give us some of the technologies we see around us. But the history of technological change in the United States is really startling in how consistently we have had inventions coming along that were of commercial value, that have managed to increase productivity, and so on. There is no evidence that this has slowed down in recent years. One is hard put to say exactly where this inventiveness came from or what kind of climate of social or government action brought it all about. There is no indication that we have seen as much as we're ever going to see of innovation either in helium extraction or helium-using technologies.

HOWLAND: The decrease in entropy, if one just calculates along the first law of thermodynamics of getting helium out of natural gas, shows that the present processes are fairly inefficient. The first law sets a lower bound as to the amount of energy one has to use to do the separation so there is room for improvement. It does not say we are ever going to get improvement in helium-extraction technologies; but we still have a long way to go before we hit the first law minimum.

McINTURFF: You cannot do better than a Carnot engine. What have we got? Approximately 1,000?

HOWLAND: It's a factor of about 800 or somewhere from 500 to 1,000.

LAVERICK: I am concerned about the remark about the socialist countries not being very advanced in research. I'm not here as a representative of the Soviet Union, but I do visit there now and again as a guest. In 1976 I wrote a report, copies of which are available to anybody who wants them, called Applied Superconductivity in the Soviet Union. It was widely circulated to my colleagues in the superconducting fraternity, at their request, with the help of Argonne National Laboratory.

decline, prices should be encouraged to rise enough to permit extraction of by-product helium and construction of new plant capacity to process leaner helium content fuel gas. Naturally conserved helium deposits and the bulk of the stored helium should be reserved until market demand exceeds these sources. Every effort should be made to forestall as long as possible the need to extract helium as a primary product from the atmosphere. Excessively high prices during the past period of abundant supply restricted use, brought regulation and legal battles. Unreasonably low prices during the coming age of scarcity will restrict production, encourage waste, and will consume our precious resources before our descendants have the proper opportunity to enjoy the fruits of our labors. If we let that happen, the judgment of history will be harsh indeed. We can do much better, and we should determine to do so.

FRANKLIN A. LONG, Cornell University: As a member of the Committee I've been impressed and encouraged by the statements so far. But I am now at the point where I would like to hear some comments from the viewpoint of the Bureau of Mines.

To the economists who have talked to us I have a particular question: Other than the legislation, is there anything negative about firm and indefinite withdrawal of Bureau supplies of helium from the market? If Bureau stores of helium were treated by legislation as a sealed-in reserve for some indefinite time, would that in any way be troublesome?

LAVE: I don't see any reason offhand why that would be bad. If the government reserves were sealed in, then the whole current market would have to come from private producers, which would certainly encourage some of these closed-down plants to open up. One can't ignore that 41 Bcf. There has to be some policy for when it's going to be sold off. It could be reserved until after the end of the century, but Mr. Howland's statements make a lot of sense. We should have a policy that maximizes the present discounted value of all that helium.

I must say that I'm somewhat suspicious of Mr. Francis' proposal since that really requires that the people at the Bureau be very farsighted indeed. Although I'm willing to believe that the people at the Bureau are intellectual superpeople, I'm not so sure that they always have a free hand in doing these things. So there is a lot to be said for cutting down the amount of discretion to be exercised and for not having them make these forecasts about what's going to happen to technology and supply and demand over the next year, but to try and state in advance a well-set-out policy so that the private producers know what to expect in future years. I am really attracted by Mr. Howland's proposal that at various intervals the government resource will be auctioned off beginning, say, in 1985 or 1990.

HOWLAND: I agree with Professor Lave. If we shut in the Cliffside field, and in the short term send all production from government-owned plants into storage, that's great up until perhaps 1987. But the essential question is: Having stored the helium, when do we begin to get a return from conservation? We need objectives and methods.

NICHOLSON: I would like to go with Dr. Long and to urge that the other side of this argument be presented. Why did we set the price as we did in the Bureau of Mines? It seems that the answer has to be clear when we consider alternatives we might pursue. I don't think from what we've heard so far that we have any idea or any defense of the action that's been taken by the Bureau of Mines over these many years.

HOWE: I'd like to ask Dr. Howland what his position is with regard to resources currently being vented? He indicated that continuing storage has no real economic role to play at present. We do have this issue of currently operational plants that are venting a resource. Marginal cost recovery is essentially zero, or very close to it. What would you recommend with regard to the allocation of this resource or the program for its use if we could ignore the potential legal implications or the legal issues that might be wrapped up in the Bureau's current attitude toward this resource?

HOWLAND: I have tried to take some possible future market clearing prices for helium and discount them back to the present at various discount rates to illustrate the effect these rates might have. Taking real discount rates of 3, 5, and even 10 percent, it seems to make economic sense to set aside the helium currently being extracted by the existing plants. If one talks about new construction of additional plants right now for the purpose of storage alone, that's another question. I would just as soon not talk about new plants in regard to your question. However, we cannot ignore the legal legacy, and I don't think that any proposal to store more helium at the taxpayers' expense is going to fly with Congress. Too many people know what the program has done so far in terms of fiscal objectives set for it in the first place. Next, so far as the legal legacy itself is concerned, one of the defenses employed by the Justice Department in fighting breach of contract suits is that we don't need to store the helium. If the Congress were to pass legislation authorizing or demanding further purchase of helium at taxpayers' expense, it would substantially increase the possibility that companies suing the government for breach of contract would win. These contingent liabilities are something like $400 million, and as economists we can argue that just to transfer money from one place to the other really doesn't have much effect on the physical allocation of resources in the economy. One can gripe about transferring money from taxpayers in general to stockholders of companies doing the suing, but to the legislators the contingent liability would tend to be a bit frightening. The administration now is scraping around for dimes and nickels and probably can't find money in the budget to authorize any new program. I would say that, on the fundamental economic merits, further storage is definitely justified. Done at the taxpayers' expense, we run into a can of worms. That's why my suggestions were all in the direction of trying to encourage private companies to store helium at their own risk and expense. It would have very little effect upon the litigation in progress, and would cost the taxpayers virtually nothing.

JOHN K. HULM, Westinghouse Research and Development Center: As I understand it, of the four plants that had contracts with the Bureau of Mines to supply helium starting around 1961, one plant is currently recovering helium and storing it in the Cliffside field but the government isn't paying for it. The helium still belongs to the company according to a report from the Bureau of Mines. In the second field the natural gases ran out, and the plant is shut down. Two other plants are still running, but the helium is being vented. In 1975 the Bureau did make arrangements to store the gas on behalf of the private owners of those plants. I'm puzzled as to why only one of the operators has taken advantage of that arrangement. The other two are still venting. We're losing a billion cubic feet each year of helium that could be stored--as each speaker has said--with no marginal economic costs. There seem to be financial or other obstacles preventing this storage. I wonder if one of you could enlighten us on the hangup here. What is the crucial question? Why don't they open the valves and let the helium go into storage?

FRANCIS: People from both Phillips and Cities Service could answer that question better than I can. However, it might be embarrassing for them to do so. As a complete outsider, I'm going to make all sorts of assumptions. I think the most critical problem of all is the nature of the various litigations that we're involved with. Litigation with respect to landowner and producer royalty payments presents a problem to Phillips in respect to helium they have produced and sold to the U.S. Government. The only decisions rendered by the courts would indicate that they are liable to pay between $12 and $17 a thousand to landowners and producers. If that same type of royalty payment were, by law and the interpretation of the courts, also placed against helium privately stored by the same plants, the financial cost--I'm not talking about economic costs--would be horrendous. Prudent management of stockholders' interests precludes taking any kind of gamble until the courts can resolve this issue. One of the crucial things the government can do is to try, insofar as it is humanly possible, to expedite a reasonable settlement of the landowner and producer royalty case.

I believe that Cities Service has somewhat the same kind of problem. Although, they're involved in a different case, the burden may not be quite so large even though they face the same kind of problem. I'd like to go back in history and put my finger on the timing of what I think is the most unfortunate circumstance. When, finally, the decision was reached by the Secretary of the Interior, after the environmental statement had been issued, to terminate storage, the various companies involved had already indicated to the government their desire to continue to enter helium physically into the storage system. Reserve the question of who owned that helium and who would have to pay for it and at what price for the completion of the litigation. Upon determination of the cases the appropriate payments would be made and ownership of the helium would be established by the decree of the courts. Unfortunately, in my estimation, the government chose not to do this and, in fact, the judge involved in the cases could not, on his own, take action to order this. The question of how this helium can be stored is a difficult one. This may take some subtle maneuvering--maybe some private agreements amongst organizations--but it seems to me that first and foremost is the issue of royalty payments. I'd like to emphasize that because it is one area that is often overlooked.

The question about the value of helium in native gas streams before extraction is something nobody has tackled with any degree of sophistication. There is no value intrinsically to helium. Helium is absolutely worthless until a user establishes the value by finding a way to put it to use and creates a demand. The value is determined by demand and demand has to be considered in a time frame. When you have a small current demand and a large current supply, it strikes me as questionable if we find high economic value for the element in the raw material state. I'd like to elaborate on this and identify some remarks that I made previously to illustrate how they are pertinent.

I made the claim that 7 of the 12 plants were constructed uselessly. One of those was the Keyes plant of the Bureau of Mines--the first of the 12 to be constructed. It was ordered and constructed in the same general time frame as the start of the conservation program. The field from which Keyes draws its gas is rich in terms of helium content. It's a better than 2 percent field, and one can

The Russians seem to be doing a lot better than we are in some areas, and it bothers me. They don't seem to have the same fight for funds. The only team in the world which has so far had courage to tackle the building of a superconducting Tokamak is the team at the Kurchatov Institute in Moscow. Without superconducting magnets, magnetic fusion containment reactor systems will be out of the question. Some U.S. scientists have criticized this Soviet work on grounds that the conductors are primitive and the industrial facilities poor. However, in spite of these disadvantages, there is a superconducting toroidal machine in existence, and it is in the Soviet Union.

In 1957 we were all somewhat smug, and we were very surprised when Sputnik was launched. However, this saved the helium program, because then we decided to go to the moon. So, thanks to the socialist societies, we were able to go to the moon and now have a justification for our helium program. However, to maintain parity, or stay ahead we should attend diligently to our research and development efforts.

The most advanced magnetohydrodynamic system in the world is on the Moscow power line. MHD in the United States has had its ups and downs. It was not mentioned in the Ray report or in Project Independence. It has suddenly been rediscovered because of the U-25 and the good showmanship of Professor Sheindlin, its director. With good common sense ERDA has formed a team to collaborate with the USSR in MHD.

In the Efremov Institute, I saw some compact superconducting switches that were about the size of a packet of cigarettes. The smallest switch, constructed from non-inductively wound, 20-micron thick niobium titanium ribbon, potted in an epoxy resin, was rated at 50 MW and had been tested at 10 kV and 2.5 kA. A slightly larger switch was rated at 80 MW. These were, and probably still are, the most advanced superconducting switches in the world. The world's largest truck factory is being built beyond the Urals, using U.S. design engineers and IBM computerized control systems. The commitment to superconducting machine development in the Soviet Union is as advanced, or probably more advanced, than anywhere else. As far as I know, General Electric, the All Union Research Institute for Electrical and Machines and Electrosilia are partners in this new technology.

Professor Gleboff who is now responsible for applied superconductivity in the USSR Academy of Sciences told me that the Soviet Union was committed to the application of superconductivity and the use of helium-based technologies and it was not going to stop until they had been fully developed. We would do well to develop a similar commitment in the U.S.

LAVE: I can see that trading barbs with Charles Laverick is very dangerous, but I can't resist. The instances you're giving are allocations of government research dollars. The clear history there is that those dollars were allocated not by economists -- since we've never had one as director of the National Science Foundation -- but by blueblooded physical scientists. There, it seems to me, the quarrel is with the very large amounts of money that went into accelerators as distinct from the amounts of money that could have been spent elsewhere.

I have no particular comment as to whether a democratic bureaucracy will be more efficient than a socialist bureaucracy. There are things to be said for socialist bureaucracies in such a case. The general remark which prompted all that was one about our research in general and its applications. There I would guess that the private sector is what I would hold out as having had the most success. Even if one thinks about government-funded research and development across the map I wonder who has the lead not just in particular technologies that one can pick out, but in general. I would guess, in spite of the very pessimistic things that are often said in the fight for more funds for the NSF, that we seem to have a good edge.

An Overview:
FUTURE REQUIREMENTS FOR HELIUM
AND SOME UNCERTAINTIES REGARDING
HELIUM SUPPLY AND DEMAND IN THE FAR FUTURE

Charles Laverick
Consultant

Introduction

The object of this discussion is to show that
technologies based on helium, in particular with regard to
energy, defense, transportation, and industry are in a
healthy state of development and not so highly speculative
as some people may suppose. Of course, the acceptance of
these technologies and their widespread application depends
upon many things: economic viability at the time of
possible application; availability of alternative options or
resources, if any; energy and resource needs of future
society; and the relative price and availability of the
helium needed. By discussing some uncertainties and
possibilities, I hope to show that if these helium-based
technologies were available in the reasonably near future,
they would meet a definite need. Some particular emphasis
has also been given to arguing that their use in space,
communications, and defense might become a paramount factor,
in view of the fact that past estimates of the appropriate
departments of government, as given to the Department of the
Interior, have been surprisingly low. These are areas where
economics, conservative design, and 30-year operational use
requirements are less important than saving space and
weight, improving efficiency, and applying the latest
technological advances as soon as they are developed.

Superconductors and Their Advantages

The various future uses of helium have been discussed
frequently in the voluminous helium literature; the concern
here is principally with nondissipating uses in
superconducting technology where helium in liquid or gaseous
phase at around 4K (-269°C or -452°F) provides the low
temperature environment at which the superconductor
transmits electrical current at very high density, even in

79

the presence of extremely high magnetic fields, without loss. In fact, a superconducting ring, closed up on itself can maintain a circulating current indefinitely as long as it is kept cold.

Because of the very low temperature involved, heat flows from the room temperature environment to boil off the helium liquid and dewar flasks, usually of stainless steel, are needed to reduce this heat leak. Refrigeration systems are used to compensate for this loss and these require some energy input. In the case of large electromagnets, a factor of 50 or 100 less power is needed for the refrigerator of a superconducting winding than is required for a water cooled copper magnet of similar performance. Where the magnets become very large or very intense, the copper magnets become impossibly expensive to energize continually and superconducting magnets become the only possible choice. Thus, in fusion or magnetohydrodynamics, the energy needed for copper coils would be more than the unit could generate, while in circular particle accelerators for high energy physics, it would be out of the question for a nation to consider the most advanced devices now under development or construction if superconducting magnets were not available; the cost would be prohibitive.

The high field properties of superconductors were first discovered in 1961, followed shortly afterward by many research magnets wound from the new materials, and by the first large practical solenoidal system to be used in a significant and expensive high energy physics experiment in 1964. The experiment itself was not carried out until two years later. In high energy physics, the Argonne National Laboratory's 16-foot diameter, 10-foot high, 1.8 tesla (T) solenoidal magnet marked the first use of such a large superconducting device in a $20 million installation. In this case, the coil provided the magnetic field for a 12-foot diameter liquid hydrogen bubble chamber, used to detect the debris from sub-atomic interactions and measure their properties. This coil has operated without trouble of any kind for more than 25,000 hours. These developments were paralleled by attempts to apply the new magnets in a wide range of engineering applications and to apply superconductors in a wide range of instruments, in computers, and in the transmission of electric power.

Superconductor Engineering in the New U.S. Accelerators

Perhaps the most spectacular and significant engineering applications are still being pioneered, because no other reasonable solution exists, in high energy physics. Two current projects involve the showpieces of the nation in high energy physics; the Fermi National Accelerator

Laboratory in Batavia, Illinois, and the Brookhaven National
Laboratory on Long Island, New York.

The 3.9-mile circumference of the main ring of the
Fermilab accelerator is clearly seen in the picture of the
site given in Figure 1. The ring contains 774 water-cooled
electromagnets with copper windings, each 20 feet long, 11
tons in weight, 1 foot high and 2.5 feet wide, to bend the
proton beam around its circular path. These are dipoles,
since the magnetic field is transverse to the magnet section
and unidirectional. The focussing magnets, 180 in all, are
7 feet long, weigh 5 tons and are spaced around the ring at
intervals of 100 feet. These are called quadrupoles, since
they have four poles and focus the beam to prevent
dispersion as it travels around the main ring. The protons
travel around the ring about 50,000 times per second,
receiving a 3 million electron volt (MeV) boost each time,
to reach final energies of 200 to 500 billion electron volts
(GeV) at a maximum magnetic field of 2.25 tesla [T] for full
energy before being used in the experimental areas as ultra
microscopic probes to study the world inside the atomic
nucleus and the forces that bind the nuclear and atomic
structures together.

Work is now underway to install a ring of about 1,000
superconducting magnets, 774 dipoles each 23 feet long and
240 quadrupoles each 4.4 feet long, below the main ring
magnet and boost the machine energy to 1,000 billion volts
(1 Teraelectron volt, 10^{12} volts), at a maximum operating
magnet field of 4.5 tesla. This maximum field can be
reached in tens of seconds. Enough of the basic 0.027 inch
multifilamentary niobium titanium superconductor, embedded
in a copper matrix of about twice the total cross section,
will be used to go once around the world. This new concept
was originally called the Energy Doubler/Saver because the
use of superconducting magnets could double the maximum
machine energy while saving $15 million per year in the
energy cost of operating the conventional main ring magnets,
by energizing these to lower fields and hence lower
currents and transferring the beam to the superconducting
ring to achieve the desired final energy   (see Figure 2).
The new proposed system has been christened the "Tevatron."
(See "The Tevatron," R.R. Wilson, Physics Today, October
1977.) A full size model section of a superconducting
dipole for this accelerator is shown in Figure 3. Inside
the beam tube a very high vacuum must be maintained to
minimize beam scattering. This is achieved by cryopumping.
The cold surface at a helium temperature of 4.6K freezes out
trace gases.

Vacuum cryopumping using liquid helium has recently
become commercially competitive for a wide range of vacuum
pumping uses and can be considered as a relatively new and

FIGURE 1   Aerial view of the accelerator, experimental areas and control laboratory at Fermilab. Hydrogen atoms, stripped of their electrons to become protons, leave the Cockcroft-Walton generator at 750,000 volts and are accelerated to 200 million volts in the linear accelerator building near the high-rise office building when they enter the small ring of the rapid cycling booster synchrotron to be accelerated to 8 billion volts (8 GeV) for entry into the 3.9 mile circumference main ring where they can be accelerated to 200 to 500 GeV before dispatch to the meson and proton experimental areas in the upper left of the picture. (Courtesy of Fermi National Accelerator Laboratory)

FIGURE 2   A view inside the main ring tunnel of the Fermilab 500 GeV proton synchrotron with the main ring magnets in the center of the picture under the service pipes and some sections of the new superconducting magnet ring in place beneath the main ring. (Courtesy of Fermilab)

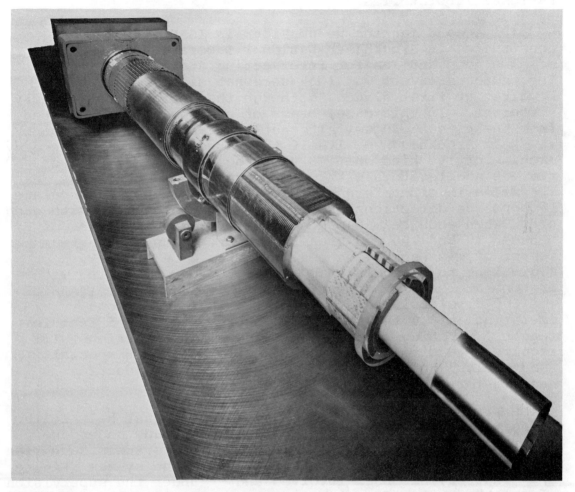

FIGURE 3   A sectioned dipole element from the supermagnet ring of the Fermilab Tevatron. The vacuum beam tube is at lower right followed at left by insulation, then the dipole coils. The magnet iron is at upper left. (Courtesy of Fermilab)

upcoming large use. Before leaving this topic, we should
consider the impact of the Tevatron on the helium liquefying
capacity of the United States. It can be seen very readily
from Table 1 that this capacity will be more than doubled.
However, the next device we consider briefly will have
similar requirements so that its completion will treble our
liquefying capacity. This is for machines in only two
national laboratories!

The second machine we consider is ISABELLE, to be
constructed in Brookhaven National Laboratory from BELLE
meaning good and ISA for Intersecting Storage Accelerator.
The proposed layout may look somewhat similar to the
Tevatron at first glance, although the devices are markedly
different. In this case, beams of protons are transported
from the present 30 GeV Alternating Gradient Synchrotron
(AGS) at Brookhaven -- itself the largest and highest energy
machine of its kind only 15 years ago -- to each of two
storage accelerators with superconducting magnet rings, each
1.6 miles in circumference, where the beams would circulate
in opposite directions and then be made to collide with each
other when each beam had been accelerated to energies of
between 30 and 400 GeV (see Figure 4). Thus the maximum
center of mass energy on collision would be 800 GeV,
equivalent to a single beam at 400,000 GeV colliding with
protons at rest. The superconducting magnets for ISABELLE
use 800 dipoles in the two rings, each 16 feet long
operating at 5.0 tesla and 276 quadrupoles each 5 feet in
length. Matching sections and experimental sections will
require an additional 84 special dipoles and 96 special
quadrupoles, to make a total of 1,300 superconducting
magnets for the rings.

Satisfactory conductors, magnets and magnet half cells
of two dipoles and one quadrupole have already undergone
realistic tests and advanced refrigeration systems have been
developed and tested in conjunction with the system (Figure
5). Industry has been involved and -- under the supervision
of Brookhaven staff, as at Fermilab -- has developed a new
technological capability. Commercially fabricated
conductors and magnets, with the necessary degree of
repeatability from conductor to conductor and magnet to
magnet will be available. Already, developments in
Brookhaven make it reasonable to consider extending magnet
fields to 6.0 tesla for these pulsed systems, so that 600
GeV beams could be accelerated and confined in rings of the
same diameter.

What of the future? Already there have been proposals
for circulating protons and antiprotons in opposite
directions in the existing Fermilab tunnel so that center of
mass collisions of 2,000 GeV could be carried. To make that
energy available by bombarding a fixed target with a moving

TABLE 1   Commercial helium liquefaction capacity in the United States
as of 1974 compared to the central helium liquefier of the Fermilab
Tevatron (1,000 GeV superconducting proton synchroton installation).

Plant	Liters/hr.
Kansas Refined Helium	700
Cities Service No. 1	700
Cities Service No. 2 (With Added Compression)	1000 (2500)
Phillips-Gardner Jayhawk	650
Linde (Amarillo)	600
Total Capacity (U.S.)	3650
Central Helium Liquefier	4000-5000

FIGURE 4 An ISABELLE magnet cell of quadrupoles and dipoles undergoing test in the experimental area. Eventually, almost 4 miles of superconducting magnets will be needed in total for the two magnet rings. (Courtesy of Brookhaven National Laboratory)

proton would require an accelerator with an energy of 2 million GeV. Such an accelerator, without superconducting technology would encompass most of the American continent, with a diameter from New York to San Francisco. Such is the power of the new technology! Yet it is only slightly over 20 years ago that the Brookhaven Cosmotron made available in the laboratory the 3 GeV particles with energy equivalent to cosmic rays. Since that time there has been an explosion in technology. What of the other laboratories and other countries? One can only say that without superconductivity, high energy physics would not have its present potential.

Superconductors in the Electric Power Industry

We turn now to electric power. Alternative national energy strategies for the future all involve increased electrification. This is well brought out in the ERDA 1975 and 1976 reports on the energy related applications of helium and their annual progress reports on energy research, development and demonstration. The idea of concentrating several high-power-capacity generating plants in one place -- an energy farm -- because of limited site availability will create a need for high-capacity long-distance electrical power transmission lines and for advanced high-capacity turbogenerators. Superconducting transmission and generation of this power to urban areas is a possibility. Similarly, the idea of mine mouth generating stations to reduce energy and financial expenditures connected with coal transportation will require enhanced long-distance electrical power transmission capability. Attempts to match daily variations in demand to minimize installed generating capacity will result in improvements in energy storage techniques. One possibility is superconducting magnetic energy storage. Thus, the tasks to be faced in increasing the electrical supply of the country are to find new energy sources and improve generating and transmission efficiency.

Some ways in which these aims can be realized involve helium and superconducting technologies. One new energy source which, if feasible and practicable, is the development of thermonuclear fusion power stations. These need helium for the magnets and several other components. One way to improve the efficiency and reduce the cost of electricity generation is to convert some of the energy in the flame of a coal-fired station or high-temperature fluid of a nuclear station into electricity before using the remainder to power a conventional steam-powered generating plant. This is magnetohydrodynamic power generation. These two processes are components of our scientific and technical exchange with the Soviet Union. They are ahead of us in these areas and have deployed substantially greater manpower and resources on these programs than we have done. Another

way to improve overall system efficiency is to attempt to
develop electrical storage capability to reduce the total
amount of spare generating capacity needed to cope with the
daily variation in load. At the moment about twice as much
installed capacity is needed to cope with overall annual
demand, i.e., half this capacity operating continuously
could supply the yearly demand. One possibility under
consideration for this storage activity is superconducting
magnetic energy storage, in which the surplus electrical
energy is stored in large superconducting coils and released
as required. Yet a third way to save space, weight, money,
and resources while improving efficiency and increasing
possible unit power generating capacity several times is to
develop power station turbogenerators with superconducting
windings.

About 13 percent of the electrical energy is lost in
transmission and distribution. In addition, we are aware of
the pressures on land use in urban areas and the demands to
improve the safety and appearance of these transmission
corridors. This can be done most efficiently -- in urban
areas requiring the largest blocks of electrical power --
and with minimum land use by using superconducting power
transmission lines. The transmission of direct current over
long distances using superconducting lines is also under
consideration. At the present time, only about 5 percent of
the total electrical power is transmitted underground,
because the cost of underground lines is about ten times
that of overhead systems. Nevertheless, in built up areas,
land costs are increasing and the need to eliminate
unsightly towers (which, in some places, interfere with
radio and T.V. communications and low flying aircraft) will
almost certainly lead to a greater use of underground A.C.
distribution lines in such areas.

The recent blackout in New York gives us some idea of
the need for better technology. Consolidated Edison has
been forced to get rid of old downtown stations and has not
been able to replace them. However, Manhattan is still the
big consumer. They cannot add more overhead transmission
space in Westchester County with the present system; more
power transmitted means higher line voltages and more space.
To enhance the capacity of existing underground lines in the
same space requires new technology. A recently completed
study for Philadelphia Electric, supported by ERDA and EPRI
at a cost of around half a million dollars considered 16
different transmission technologies with the same
guidelines. The Brookhaven Laboratory's superconducting
line study was one of the lowest cost technologies
identified and thus superconducting lines are a viable
option. The three of four front runners were only a factor
of 3 or 4 higher than overhead lines in capital cost.

The Brookhaven 300-foot-long test line is shown in
Figure 6. There is only a pipe to see, but its current
carrying capacity is phenomenal. It has to be compared with
the batteries of lines with which we are familiar. This
line has been designed for utility type service and the pipe
has been optimized for commercial production. The flexible
superconducting cable inside the pipe is rated at 138 kV, 6
kA; a capacity of 1400 MVA. This exceeds the capacity of
the new Shoreham 800 MWe nuclear plant being built on Long
Island. The refrigerator for this demonstration line has
been designed specifically as a prototype for refrigerators
intended for electric utility company service in advanced
power transmission systems. High reliability has been
achieved by the use of rotary components; the compressor is
an oil-injected screw machine rated at 350 horsepower and
expansion is achieved by means of three gas-bearing
turbines.

The specification and overall assembly were the
responsibility of Brookhaven; the major components were made
by Cryogenic Technologies Inc., Waltham, Mass.; Dunham-Bush
Inc., Hartford, Conn.; and Sulzer Bros. Ltd. of Switzerland.
The system incorporates many state-of-the-art advances. For
example, oil carryover to the coldbox is continuously
monitored by means of a laser reflectometer.

After several days of last minute checking by Brookhaven
staff members and representatives of the major component
manufacturers, the machine was started on the afternoon of
May 19, 1977. Within 12 hours the refrigerator achieved the
design goal of 500 W at an average output temperature of 7
Kelvin. The refrigerator does not produce liquid helium.
The helium is in the supercritical regime at a pressure of
15 atms. In early June during a 48 hour continuous run, a
total heat rejection of 750 W was obtained under the same
output flow conditions.

The work is part of the Power Transmission Project which
is funded by the U.S. Department of Energy. The design of
highly reliable, efficient helium refrigerators is also a
crucial element of many other DOE programs including fusion
energy development and high-energy particle acceleration.

Superconducting Magnetic Energy Storage also deserves
mention at this point. A small component of DOE's activity,
the studies are being carried out at Los Alamos Scientific
Laboratory and the University of Wisconsin. Conceived for
load leveling, these devices would require little space and
improve overall system efficiency, thereby reducing the need
for resources and land space for the added central station
power, eliminating some environmental degradation and
reducing water needs. Thus, it might become attractive
where further environmental degradation is intolerable or

FIGURE 5 An artist's conception of the ISABELLE magnet rings in the main tunnel. (Courtesy of Brookhaven National Laboratory)

FIGURE 6 View of the 300-foot demonstration superconducting underground electrical power transmission line installed in Brookhaven National Laboratory. The flexible superconducting cable inside the pipe has a rated capacity of 1,400 MVA, much in excess of the largest existing nuclear power plant, and its helium refrigerator has been designed specifically as a prototype for electric utility service. (Courtesy of Brookhaven National Laboratory)

where water resources become critically short or valuable. The program calls for Los Alamos to help the Bonneville Power Authority design and build a 30-megajoule (about 8 kilowatt hours) power system stabilizing magnet by about 1979, to be followed around 1985 by a 3 to 10 megawatt hour stabilizing unit if a suitable utility site can be found, with a 5,000 megawatt hour demonstration system to be completed at some utility site around 1990. Such a unit would require 6 million liters of liquid helium inventory (about 150 MMcf). It would have a discharge time of 5 to 10 hours so that it could act to level the load on a large power system and may be competitive with hydrostorage in many areas. The concept is shown in Figure 7.

Yet another concept of interest to the United States is magnetohydrodynamic power generation (MHD). Although many enthusiasts have carried on development over the years (e.g., Avco Everett Research Laboratory) without much support, it has been viewed with suspicion as being too difficult or as not promising to be economic. Claims of increases in efficiency of 10 to 15 percent have been made for MHD peaking units. On a 2,000 GWe national installation -- as has been projected for 2000 AD in some estimates at an average of 33 percent efficiency -- the input energy resource has to be around 6,000 GW thermal. A savings of 15 percent brings the input resource to around 4,000 GW thermal. This extreme example serves to show the order of magnitude of the possible resources involved and the potential for saving resources, hence capital and environmental degradation. Even if placed on one in five plants, MHD seems to represent a worthwhile subject for study. A U.S. program is now underway. The U-25 plant in Moscow has already achieved worthwhile successes; notably, it has operated at more than 20 MW and continuously at lower power for more than 100 hours. A channel life of 3,000 to 4,000 hours is desired (about 6 months continuous operation) to make the concept attractive.

Since it will be some years before the United States has a comparable installation, U.S. teams are hoping to gain experience on the U-25 MHD plant in Moscow, USSR. As an incentive to accept such a collaboration and to examine the performance of superconducting magnets in a working situation, a U.S. designed magnet has been installed in a bypass channel on the U-25 and has been operated "in-situ." Built in Argonne National Laboratory in the previous year, it was successfully tested to design field (5 tesla) in April 1977, then flown mounted on its own truck to Moscow in a U.S. Air Force C-5A Galaxy on June 15 to make a spectacular debut for the beginning of the World Conference on the Future of Electrotechnology and the associated exhibition Electro 77. It was driven from the C-5A and installed the same day. The preparation for departure

91

FIGURE 7 Artist's conception of a superconducting solenoidal energy storage unit for load levelling on a large power system. The 5,000 mega-watt hour system uses a magnet supported by bedrock. As shown, the system uses superconducting d.c. power transmission lines for maximum efficiency and minimum environmental effect. (Courtesy of the Department of Energy)

(Figure 8) from O'Hare Field, Chicago, and subsequent installation in Moscow (Figure 9) emphasizes the importance of superconductivity and its associated helium to this endeavor. The completed coil is shown in Figure 10 before final bonding and insertion into the helium container and associated vacuum vessel. Its principal components and dimensions are given in Figure 11.

The success of this magnet has led to the development of a similar coil for our own US-MHD program. This will have a slightly higher magnetic field, 6 T instead of 5 T and will be used in a Component Development and Integration Facility (CDIF) now under construction. The system should be ready for test in 1978 using a conventional iron core electromagnet operating at a maximum field of 3 T. Operational test conditions with coal will be achieved and various MHD components at the 50 MW thermal level will be tested and integrated with other components on a prepilot scale. This facility will incorporate a high field test bay with a superconducting coil which is estimated to be ready about one year later, say by the beginning of 1980. This system will be followed by a 250 MW thermal Engineering Test Facility (ETF) with a tentative completion date of around 1984. It will use a superconducting coil with an inlet field strength of 6 T in a 3-foot-bore inlet diameter, an outlet field strength of 4 T in a 6-foot-diameter and an active duct length of 23 feet.

The final step in the program will be the development and construction of a Base Load MHD plant on a 1,000 MW scale for commercial feasibility demonstration around early 1990. The Soviet Program is expected to have such a plant on line by 1985. Design studies are underway. The impetus to provide a superconducting magnet for the CDIF facility by about 1980 clearly has received a boost from the Soviet initiative, our willingness and ability through pioneering efforts to build a successful superconducting dipole for the U-25 line and the good sense of both countries to arrange a collaborative program through which each serves his best interests by providing a joint team from both countries to work on the U-25 and its superconducting dipole in the bypass loop.

As presently envisaged, the Base Load MHD plant would be a peaking plant developing about 50 percent of the total electrical output followed by a conventional steam plant. The tapered bore superconducting magnet for such a plant would have a field strength of 6 T in an inlet bore of 7.4 feet, a field strength of 3.5 T at the outlet bore of 15.6 feet, an active duct length of 52.5 feet and an overall length around 82 feet. These magnets will become a reality and are not trivial engineering tasks. Such a coil will weigh 2,500 tonnes (1 tonne = 1,000 kilograms = 2,200 lbs),

FIGURE 8   The U.S. 5.0 tesla superconducting MHD dipole magnet system on its transporter being loaded into a U.S. Air Force C-5A Galaxy at O'Hare Field in Chicago for delivery to Moscow Airport, USSR. The million-dollar system was constructed in Argonne National Laboratory and successfully operated to its designed operating field during its first test in April 1977. (Courtesy of Argonne National Laboratory)

FIGURE 9   The U.S. superconducting MHD dipole being installed in its place in the bypass beam line of the USSR 20 Megawatt U-25 MHD system at its location in the High Temperature Institute in the Moscow suburbs,   USSR, on the afternoon of its arrival at Moscow Airport. (Courtesy of Argonne National Laboratory)

FIGURE 10 An inside saddle winding of the U.S. superconducting MHD system, to show the configuration of the composite superconducting saddle winding, insulating fillers, spacers for liquid helium flow and reinforcing support rings. (Courtesy of Argonne National Laboratory)

require 500 tonnes of conductor of which 50 tonnes will be niobium titanium and the remainder copper stabilizer. Currents of 100,000 amperes in the conductor open the door to a new era of very high current technology. The total transverse magnet load will be 100,000 tonnes, requiring a substantial containing structure to withstand such loads. The problem of reducing helium inventory in these coils has received much consideration. If the ring girders were located within the helium vessel, a liquid helium inventory of 2 million liters would be needed, equivalent to 49.4 million standard cubic feet of helium, at a cost of 20 percent of the magnet or $6 million in 1976 dollars. Using the need to minimize helium use as a serious design constraint, attempts are being made to reduce this inventory to 20,000 liters (50,000 cubic feet of helium, i.e., 50 Mcf in gas industry terminology). Thus 500 such units, a not unreasonable number, would need at minimum 25 MMcf of helium with a makeup of 2.5 MMcf per year at 10 percent loss rate.

We turn now to fusion. A key element in the magnetic fusion energy program is the Tokamak machine. This is a toroidal device in which the reacting deuterium-tritium plasma is held together in a doughnut shaped vacuum vessel by a ring of superconducting magnets arranged around the doughnut. Between the magnet coils and the vacuum vessel, a neutron absorbing blanket is needed to capture the 14 MeV neutrons generated in the reacting plasma as the ions fuse together to form heavier elements. This blanket could be helium cooled to remove the heat energy and transfer it, by means of a heat exchanger, to produce the steam needed to drive the turbogenerators. The turbogenerators themselves could have superconducting windings and, in fact, helium gas turbines and MHD peaking units are not out of the question.

Helium is required in large amounts for the cryopumps to maintain the necessary low gas pressure in the vacuum region, and either as a liquid or two-phase fluid at or near 4.2°K for the superconducting magnets. In the ERDA-13 report, the helium estimate for cryopumping was not considered. However, the inventory for magnet and blanket cooling was considered and was given as 11.4 Mcf/MWe for a 2,030 MWe Tokamak. This led to an estimate of 19.2 Gcf for the cumulative inventory to 2020 assuming 3000 installed gigawatts by the year 2020, a considerably lower figure than that in earlier reports [e.g., the Ray report]. It should be noted that an annual leak rate of 10 percent per year, not an unreasonable figure, would lead to a prohibitive 2 Gcf makeup requirement/year. Clearly this needs to be reduced and would be a priority experimental and design problem. Attempts are underway to carry out early scoping studies of reactor designs to highlight critical problems on the path to commercial demonstration of fusion by the 1990s. Of course, other approaches to fusion exist. All need helium. However, the Tokamak is today's frontrunner and was chosen as an example in this narrative for that reason. Attempts are underway in conceptual designs to reduce Tokamak size and overall reactor cost. However, definitive results for the principal parameters in question are not expected to be forthcoming before around 1982. As with power stations -- with the difference that all is supposed to be known when power stations are built -- the time from concept through design, construction, operation, and interpretation of results for these increasingly large devices is about eight to ten years. Since there are only 22 years to the turn of the century and the devices are increasingly expensive, we don't have too many attempts. Perhaps a Fermilab approach is needed to focus available resources on the next generation of machines. However, it is clear that time is short and since no superconducting Tokamaks have yet been built in this program, superconducting magnet experience has to be obtained soon as it is out of the question to conceive of a commercial device without such elements. Thus helium will be needed for the tests and devices of the next 20 years and in increasingly large amounts as commercial feasibility is demonstrated.

The superconducting Soviet Tokamak T-7 is already assembled and in the final checking and adjusting stages prior to test. The cost of the Magnetic Fusion Energy Program from the present to commercial demonstration is estimated at between $14 and $22 billion, depending on the time needed and the degree of priority given to the program. Without helium, or with very high cost helium, this investment and that already made, would be undercut. A typical Tokamak magnet concept is shown in the 1976 Argonne Experimental Power Reactor (EPR) reference design (Figure 12) where the toroidal and poloidal magnet coils are clearly labeled.

An interesting new Tokamak idea due to J. Powell of Brookhaven National Laboratory is shown in Figure 13. Known as DEALS for Demountably Anchored Low Stress Magnet System and the subject of a modest DOE study contract, it is based on the idea that if these systems are to operate for 30 years, they should be designed to be taken apart to provide easy access to the inside components of the machine, have demountable superconducting magnet windings and be capable of operating at the highest fields required by the plasma physics. It shows promise of being a feasible approach to future device design.

Powell, in collaboration with G. Danby, a Brookhaven Laboratory colleague, was also the initiator of another idea which has stirred great interest. Not of current interest to the U.S. Department of Transportation, several versions of the high speed magnetically levitated train have been

97

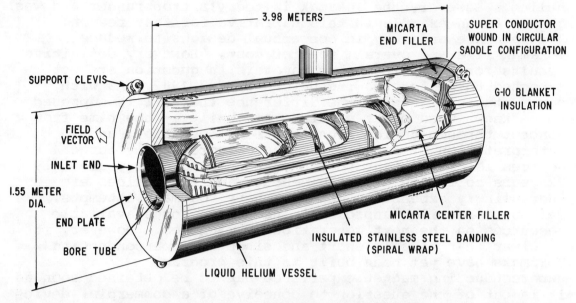

US SCMS COLD MASS

FIGURE 11   Schematic view of the low temperature section of the U.S. 5.0 tesla, U-25 bypass MHD magnet showing details of its assembly. (Courtesy of Argonne National Laboratory)

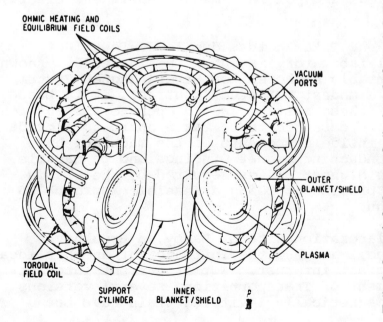

FIGURE 12   Schematic view of the magnet assembly of a conceptual experimental fusion reactor giving details of the principal components. The reacting plasma is enclosed in a vacuum vessel whose vacuum is maintained by helium cryopumps and compressed by the magnetic field generated by the toroidal field coil. The 14 MeV neutrons from the plasma are captured in the blanket where they give up their energy as heat. The blanket is cooled by a circulating fluid, helium is one possibility, which transfers the heat to a normal steam-electric plant. The toroidal field coils are at or near boiling liquid helium temperature (4.2°K). This design, due to Argonne National Laboratory (December 1976) is typical of several which have been developed in U.S. laboratories. The huge scale contemplated for these devices can be seen by contrasting its size with the standard six-foot-high man at bottom right.

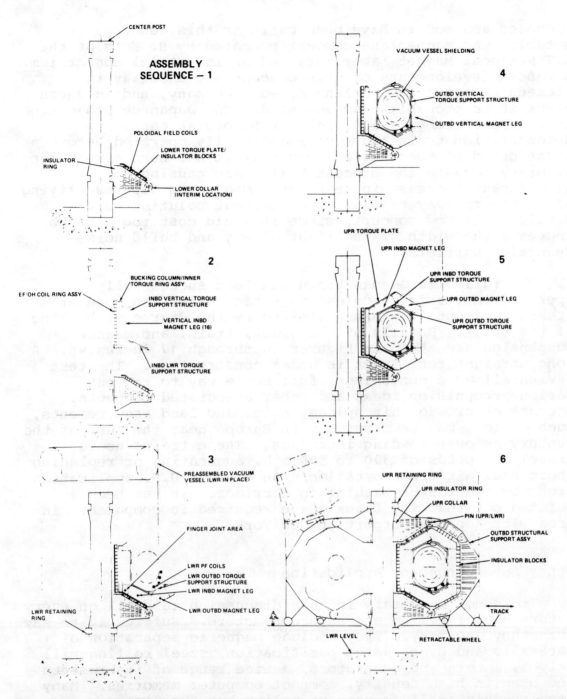

FIGURE 13   Conceptual design of a toroidal Tokamak reactor due to a combined study team from Brookhaven National Laboratory and Grumman Aerospace Corporation. Reasoning that such complicated devices, which should have operating lives of about 30 years, should be demountable this concept proposes a demountable magnet whose windings can be simply manufactured as straight sections then assembled or taken apart on site as required to provide access to all inner components. A demountable room temperature support structure, capable of acting as a reactor containment vessel, is also used. The system is designed to incorporate the brittle niobium tin superconductors required for the highest currently available magnetic fields and operate them at low stress. Such high fields are deemed essential to the success of fusion by several plasma physics groups. The concept is known as DEALS (Demountable Externally Anchored Low Stress Magnet System). The six assembly sequences shown in the figure illustrate how the system could be taken apart or put together while the final assembly sequence No. 6 depicts the completed system.

99

proposed and models have been built in this country,
notably, the Magneplane concept promoted by H. Kolm of the
MIT National Magnet Laboratory and an industrial consortium.
Advanced developments of this concept are underway in
Siemens Laboratories, Erlangen, West Germany, and in Japan
where it is probably most advanced. The Japanese project
lost support temporarily, but not before an early
demonstration train had been successfully operated before a
large crowd of spectators. Recently it has been given high
priority because the present trains are causing a
significant increase in deafness among young Japanese living
along the railway tracks. The cheapest solution is to
develop this new concept, since it would cost too much to
increase the width of the right of way and build noise
reduction barriers.

The full-size German train has been successfully
operated in the frictionless magnetic levitation mode at
speeds of about 120 km/hr, limited by the degree of banking
of the track. Details of the track, train, and magnet
suspension are shown in Figures 14 through 17 inclusive. A
long straight test track is under construction. The test
device allows a convenient, full scale way to try out
various propulsion ideas and other associated concepts.
Because of crowded air spaces, noise and land requirements,
such devices may be necessary in Europe near the turn of the
century as overcrowding increases. These trains could
travel at speeds of 300 to 500 mph, augmenting or replacing
short haul air transportation. An obvious U.S. application
might be the Boston-Washington corridor. It has been
pointed out that the track space required is comparable in
area with Kennedy Airport in New York.

Other Superconductor Applications

It is not necessary in this discussion to dwell on every
future use of helium and superconductors. Suffice it to say
that they are many. They include magnetic separation of
materials and ores, water purification, steel rolling mill
drives, motors and generators, a wide range of instruments
and uses in high density, compact computer memories. Many
require only small amounts of helium but they may require
much more in total than we might now think realistic. In my
helium study reports to the N.S.F., I estimated that there
were three other advanced aggregates of our contemporary
civilization. Europe, USSR, and Eurasia. Thus world demand
for helium could approach four times that for the United
States by the turn of the century. If China were included
in this category, since they already make superconductors
and supermagnets, world demand might be five times that of
the United States by that time.

FIGURE 14   General view of Erlangen, West Germany, showing Siemens Research Laboratories in the center of the picture and the precision circular concrete test track for testing full-scale magnetically levitated high speed train coverages at speeds up to 120 kilometers per hour. (Courtesy of G. Bogner, Siemens Research Laboratory)

FIGURE 15   A full scale test carriage of the AEG-BBC-Siemens consortium on its banked track. Supported by the West German Government, these high speed, frictionless trains of the future may supplant or replace short-haul air traffic in Europe, traveling at speeds of 300 to 500 miles per hour. (Courtesy of Dr. G. Bogner, Siemens Research Laboratories)

FIGURE 16 The banked track for the West German high speed, magnetically levitated train showing the vertical and horizontal aluminum guides. As the carriage moves at high speed, eddy currents are induced in the aluminum sheets by superconducting magnets on board. The repulsive reaction causes the train to float on the magnetic field between the horizontal plates and the four coils at each corner of the carriage, while a similar reaction between four vertically placed coils and the vertical aluminum plates constrains the carriage between the tracks.

FIGURE 17 One of the eight superconducting guide magnets on each high speed carriage. The enclosure contains superconducting coils, liquid helium dewar and vacuum insulation. The remarkably small size of the coil system needed to levitate and guide such a large carriage is surprising to many.

* * * *

Helium Technologies in Communications, Space, and Defense

Introduction

In common with all the other technological revolutions facing mankind, those concerned with communications, computation, and space are going to have a drastic impact on our lives. The present intense activity in space will accelerate. A background temperature of 3°K with a more perfect vacuum than we have on earth will favor many superconducting applications; this is the environment of space. Control of military strategic and tactical operations from space is already part of the balance of terror (Stores, 1976; Stockholm Int. Peace Research Institute, 1975). The enormous pressures on the governments of each superpower to strive for technological superiority will cause activities in space to increase as quickly as national budgets will permit (Stockholm Int. Peace Research Institute, 1975; Murray and Davies, 1976).

It seems that more and more men and robots will live and work in space (Murray and Davies, 1976) and that war is inevitable (Durant, W. and A., 1968). There will be the need to protect these stations and their inhabitants and a desire to attack, destroy, or otherwise neutralize those controlled by other groups. There will be a need for magnetic screening from solar radiation and the electromagnetic pulses from thermonuclear blasts. There seems to be a desire to develop high power laser weapons (The Future Revised, 1976). Attack satellites might become common (Murray and Davies, 1976).

Scale of the Effort

Defense is a vast undertaking. The U.S. has over 500 naval vessels of which 15 are attack carriers and over 100 are nuclear submarines equipped with the most sophisticated technological devices we can conceive (Zumwalt, 1976). Infrared laser ranging and artillery permit large numbers of our tanks and aircraft to search and destroy as well by night as by day (The Future Revised, 1976). Satellites can assess distances within feet and permit simultaneous control and communication over large numbers of different channels to sea, land, air, and other space units. Our air forces comprise almost 980,000 personnel with 13,000 planes, including many of which carry nuclear weapons at supersonic speeds in level flight while equipped with the most sophisticated devices for control, communication, navigation, and ranging (Stockholm Int. Peace Research

103

Institute, 1975). An attack carrier like the _Nimitz_ is propelled at high speed by engines of more than a quarter of a million horsepower (Stockholm Int. Peace Research Institute, 1975). Antisubmarine ships have 150,000 h.p. engines, and the nuclear submarines need engines in excess of 40,000 h.p., with minimum weight and sound and maximum cruise range. We have to contend with over 10,000 surface-to-air attack missiles, and 2,200 long-range missiles. The military applications of satellites and space will no doubt develop even further.

Airborne Refrigerators and Generators

This vast defense enterprise is budgeted at over $100 billion in 1977. Even assuming no escalation and today's dollars, this would be about $2,500 billion to the end of the century, with perhaps another $100 billion for peaceful applications in general science and space (e.g., $100 G for 25 years). Most of the equipment has to be changed every decade. ERDA feels that civilian applications for energy technology alone will require between 120 and 180 Bcf of helium to 2030, even assuming the most conservative demand projection of the Bureau of Mines to 2000 A.D. (U.S. ERDA, 1975). By comparison, this vast and complicated enterprise in defense and space is estimated to need 60 MMcf of helium per year to 2000 A.D. (at a cost of about $1.2 million) for DOD with a similar estimate for NASA (U.S. Dept. of the Interior, Bureau of Mines, 1972). No estimates for the years beyond 2000 A.D. have been published; at least, in the open literature.

The quantities estimated seem unbelievably small and take no account of the vast nature of the military-space-industrial enterprise nor of the fact that modern war involves everybody and all our resources and technology.

Let us take one simple example and assume we are fighting for our existence with 20 minutes warning to try and save 100 million of our people from annihilation. We have one laser under consideration which needs 170 Mcf of helium for a 15 second pulse which might destroy a missile. Assume it needs 10 shots to do so and there are 2,000 incoming missiles. The 2,000 shots in a short space of time might need at least 200 such lasers and the total burst would need 3.4 Bcf of helium assuming no other tests or uses. Development, test, and routine checking of such devices on this scale would need considerably more. Masers and lasers have many other uses in space and defense apart from the examples given here. Helium will also be needed for these purposes as well as for the wide range of superconducting instrumentation.

104

Many helium refrigeration systems are already used routinely in flight by the Air Force. The 5 MVA Westinghouse superconducting airborne A.C. generator rotor is accepted as a success (Parker, 1975). A reasonable estimate of aircraft numbers through 1990 would be about 3,000. Some reasonable number of new aircraft might be fitted with new lightweight high performance generators after 1980. Because of their size and weight, refrigerators will not be carried in operation in most aircraft, but may be used during standby on the ground. If we assume a basic inventory of 500 liquid liters per generator, this would imply a large total inventory. Loss rates will depend upon basic loss on stand by, plus dissipation in flight. If we assume 1,000 flying hours/years/plane at about 40 liters/hour loss, each plane would need about 1 MMcf/year. Such a use over the next 50 years would require substantial quantities of helium. Naval and civil aircraft might also take advantage of such equipment. We can conclude that a significant demand for such a purpose will develop. Superconducting airborne instruments will also need helium and high capacity superconducting energy storage concepts, for airborne and army use are also under consideration.

MHD Applications

The military establishment has had a long time interest in improving the capacity, portability, and efficiency of its power supplies. The advantages of practical MHD power generation have become self-evident. MHD units can handle very high temperatures and power levels, have no moving parts or close tolerances, and can start up and reach full power rapidly. Military applications usually precede civil applications, since economics usually takes second place to other qualities such as performance and serviceability under extreme conditions of environment and use. Where short duration bursts of power are required, the units can be more compact and lighter than most. However, as with the larger commercial units, superconducting magnets are needed, and in this case the ultimate in performance and weight reduction is needed. DOD has already sponsored the construction of 20 to 30 MW units (Cooper, 1974). Their advantages are virtually unlimited single unit rating, one to five second startup, millisecond response to load variation, low capital cost, and very low pollution. The Air Force program aims at the development of a compact, lightweight, 10 MW system. Following this, the aim is to provide continuous operation for 1 hour at $30-40/kW, and 10 seconds at $24-35/kW, followed by a 2,000 MW system at a lower per kilowatt cost. Pulsed systems are required which will develop 10 MJ in 2 msec bursts, followed by later systems capable of developing 300 MJ 2 msec pulses on a routine basis. The success of this program will lead to wide service application of continuously operating and pulsed systems for airborne, marine, and land use. Helium demands for such applications cannot as yet be assessed, but they will certainly materialize at the turn of the century.

Superconducting Electric Ship Drives

Naval applications of superconducting motors and generators are under development. The International Research and Development Company (IRD) has pioneered the development and application of homopolar motors and generators (Appleton, 1975). Based on recent tests, earlier successes, and several years' development and operating experience, they offer such devices at output ratings up to 60 MW incorporating advanced and unpublished technology (Appleton, 1974). Smaller 3,000 h.p. devices for the U.S. Navy's hydrofoil vessels are under development at General Electric as part of the general program toward the development of 40,000 h.p. units (Akerman, et al., 1977).

A 1 MW superconducting D.C. generator and 1,350 h.p. superconducting motor system for ship drive experiments was successfully tested at IRD in early 1976, and more advanced power systems are under development. The advantages of such systems for naval vessels are (Appleton, et al., 1977):

1. The prime mover has an extended life factor.

2. Different types of prime mover can be combined in the same propulsion system, and all used to maximum power simultaneously, if required.

3. The superconducting system can be arranged to operate with maximum fuel economy at all times.

4. The superconducting systems permit flexible layout of the components, with low weight and low volume. This promises radical changes and improvements in ship design.

References

Helium Technologies in Communications, Space, and Defense

Akerman, R.A., R.I Rhodenizer, and C.O. Ward. 1977. A superconducting field winding subsystem for a 3,000 h.p. homopolar motor. IEEE Trans. on Magnetics, MAG-13(1) (January).

Appleton, A.D. 1974. IRD. Private communication.

Appleton, A.D. 1975. Superconducting D.C. machines: Concerning mainly civil marine propulsion but with mention of industrial applications. Invited Review Paper J8. Proceedings of the 1974 Applied Superconductivity Conference. In IEEE Trans. on Magnetics, MAG-11(2):633, et seq. (March 1975).

Appleton, A.D., T.C. Bartram, R. Potts, and R.W. Watts. 1977. Superconducting D.C. machines - A1MW propulsion system - Studies for commercial ship propulsion. In IEEE Trans. on Magnetics, MAS-13(1) (January 1977).

Cooper, R. 1974. Presentation at the USAED-DCTR Power Supply and Energy Storage Review Meeting, March 5-7, 1974. USAF Program. USAEC, Washington, 1310. pp. 207-245.

Durant, W. and A. 1968. The lessons of history. Simon and Schuster, New York. 116 p.

The future revised. 1976. Wall Street Journal (April 15, 1976).

Murray, B. and M.E. Davies. 1976. Detente in space. Science 192(4244) (June 11, 1976).

Parker, J.H. et al. 1975. A high speed superconducting rotor. Westinghouse Research Paper J9. Proceedings of the 1974 Applied Superconductivity Conference. In IEEE Trans. on Magnetics, MAG-11(2):640 (March 1975).

Stockholm International Peace Research Institute. 1975. World armaments and disarmament: SIPRI Yearbook 1975. MIT Press.

Stores, John. 1976. The strategic nuclear arms race. Impact of science on society 28(1/2).

U.S. Department of the Interior, Bureau of Mines. 1972. Environmental impact statement, termination of helium purchase contracts, DES 72-6. Draft May 16, 1972, final November 1972.

U.S. Energy Research and Development Administration. 1975. The energy related applications of helium (ERDA-13). E.F. Hammel [Dir.]. A report to the President and the Congress of the United States. Office of the Assistant Administrator for Conservation, U.S. Energy Research and Development Administration, Washington, D.C. 112 pp. plus appendixes.

Zumwalt, E. 1976. On watch. Quadrangle Press. Chicago.

* * * *

Some Comments on Supply and Demand

Helium Availability after 2000 A.D.

Helium availability beyond the turn of the century depends a great deal on how the government stockpile is managed between now and then. With unwise management it is possible to destroy the private industry which has developed in the last 15 years, dissipate the stockpile by about 2000 A.D., invalidate a possible $20 to $40 billion investment in the development of helium-based technologies, severely limit the energy and defense strategies available to the country, hence threatening its existence and in all probability leaving it with only very expensive helium from air. (The NSF study assumed helium from air to cost between $1,000 to $3,000 per Mcf and require 26 GW(e) years per Gcf of helium extracted; the ERDA-13 study used $3,000 to $6,000 per Mcf for the extraction cost. Both studies estimated that by-product helium from oxygen extraction plants, assuming normal growth of that industry, might cost around $500 per Mcf but could supply only a small fraction of the demand, say 15 percent if used.)

With best management, assuming the storage of an additional 6 to 10 Gcf of helium by 1984 to 1987 (the 1974 figures were 18 Gcf, thus 8 to 12 Gcf of that then available has already been lost) no dissipation of the 42 Gcf in the Cliffside field and 16 Gcf in shut-in fields, we could begin the new century with 64 to 68 Gcf in hand plus whatever resource might then be available. In 0.1 percent and 0.2 percent helium-rich gas streams this could be between zero

108

and 40 Gcf, using the Hubbert natural gas estimates, to as high as 362 Gcf using the most optimistic USGS estimate.

The situation with regard to supply projections is best seen from Figure 18 where the annual helium supply for future years has been derived from estimates of natural gas available in the United States according to the USGS and M. King Hubbert, using the Potential Gas Committee estimate for the helium concentration in the gases for various geographical regions of the United States. We can see that, if the optimistic projection is correct, there is no serious problem, although it does seem improvident to throw away the cheap helium from the unique helium-rich fields when we already have operating plants for which we have paid, through which these gas streams are flowing. However, if we make our plans on the basis of these optimistic estimates and the projections based on M. King Hubbert are close to reality, we foreclose our options. In the absence of more definite information, it would seem prudent to shape our helium management policies on the assumption that the lowest estimates are correct (see Table 2). Effectively, the lower estimate leads us to the approximate conclusion that if we needed somewhere between 75 and 140 Gcf of helium from gas fields containing 0.1 percent or more to support our technological options for 30 years after their widespread application has become assured, around the turn of the century we would need to have somewhere between 60 and 125 Gcf of helium in hand, as only about 15 Gcf of resource would exist, or we might have to obtain it from air. In that period we might obtain 0.3 to 0.4 Gcf per year from the oxygen industry at $500 per Mcf in today's dollars; the remainder might cost somewhere between $3,000 and $6,000 in 1975 dollars according to ERDA-13.

Other uncertainties in forecasting the growth of helium-based-technologies clearly depend upon many national and global societal factors such as the ratio of electrical to nonelectrical energy uses, the relative desirability of helium-based technologies, population growth, energy use per capita, efficiency of energy use, worldwide supply-demand pressures, and so on.

Helium Demand and Uncertainty Beyond 2000 A.D.

Demand beyond the turn of the century is highly speculative. If it has to rely on helium from air and by-product helium from the oxygen industry, it might be quite low and drive the new technologies out of the marketplace on the basis of cost alone, depending on alternative source energy prices. If it were to expand at a modest percentage, in step with the general economy, it might be supply limited. Taking a practical and reasonably conservative

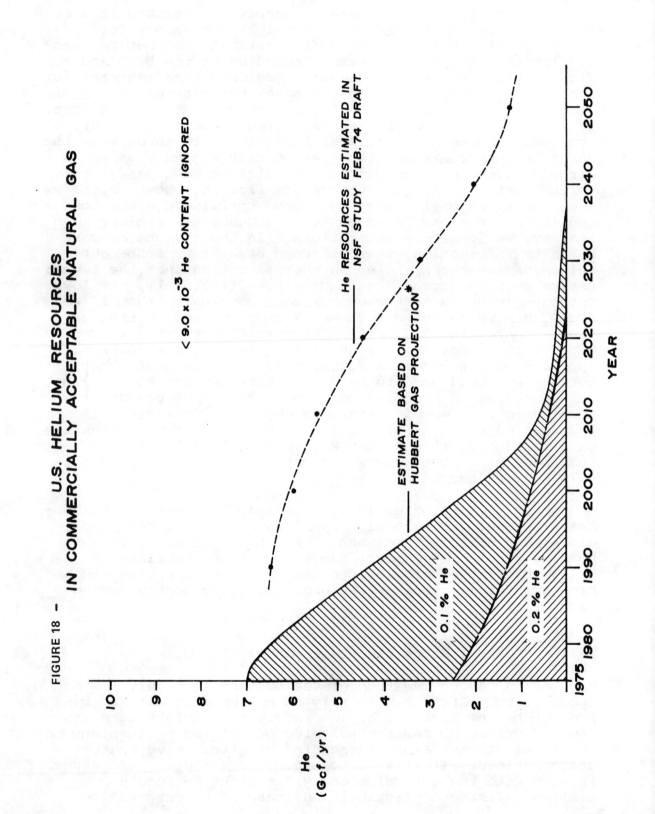

FIGURE 18 – U.S. HELIUM RESOURCES
IN COMMERCIALLY ACCEPTABLE NATURAL GAS

< 9.0 x 10⁻³ He CONTENT IGNORED

He RESOURCES ESTIMATED IN
NSF STUDY FEB. 74 DRAFT

ESTIMATE BASED ON
HUBBERT GAS PROJECTION*

0.1 % He

0.2 % He

He
(Gcf/yr)

YEAR

TABLE 2   U.S. helium resource estimate based on Hubbert estimate of gas
resources and the Potential Gas Committee estimate of helium concentration
in various regions.

Year	1980	1994	2000	2010	2020	2030
Total Gas Tcf	23	12	8	4	2	1
50% Processed Tcf	11.5	6	4	2	1	0.5
0.2% He Content Tcf (9%)	1	0.5	0.4	0.2	0.1	0.05
0.1% He Content Tcf (40%)	4.6	2.4	1.6	0.08	0.04	0.02
0.2% He Gcf	2	1.0	0.8	0.4	0.2	0.1
0.1% He Gcf	4.6	2.4	1.6	0.08	0.04	0.02
TOTAL HELIUM Gcf	6.6	3.4	2.4	0.48	0.24	0.12

view, as was done in the NSF report, one might merely take the Bureau of Mines median estimate of 2.5 Gcf per annum, as given in the 1969 hearings, and either extend it at a constant annual consumption ratio (for the U.S. only) or allow it to increase at the modest percentage rates (1 percent to 3.5 percent) assumed in the Ford Foundation Report (Ford Foundation, 1974). It should be noted that if we accept the 1 Gcf helium market sales for 1977 and 2.5 Gcf for 2000 A.D., this would give a not unreasonable growth projection of 4 percent per annum over the 23-year period. Thus in the 30-year period from 2000 A.D. to 2030 A.D., assuming 2.5 Gcf per year in 2000 A.D., we would have a total demand of 75 Gcf at zero growth, 88 Gcf at 1 percent per annum, and 140 Gcf at 3.5 percent per annum.

The Future U.S. Energy System and Its Uncertainties

The division of energy use in the 1972 U.S. reference system is shown in Table 3. Out of 72.2Q, 50Q was used directly and 22Q was converted to electricity, from which users received 5.7Q of electrical supply. Thus 16.3Q was lost in the conversion, an efficiency of 26 percent. Clearly, superconducting technologies offer an opportunity to conserve energy and resources in the electrical center since they offer increased efficiencies and a decrease in the amount of resources consumed for a device of given performance, and they are usually smaller and lighter than equivalent nonsuperconducting devices.

As seen two years ago, the probable components of the U.S. energy system in 2000 A.D., together with take-off dates and potential contribution, are given in Table 4. Superconducting technologies may be involved in some of the solar components or in transmission, but this is not clear. The part they would play in enhancing 2000 A.D. energy supply with conservation practices is also not clear (see Table 5) at least according to government estimates.

Another factor to be considered in assessing future energy needs and the possibilities for future technologies is the size of national population at the time considered and the energy per capita use at that time. World population is also important, since it could exert pressures on the advanced countries by limiting the extra-national resources available to them. The uncertainty of the national problem is seen in Table 6 and Figure 19. Table 6, developed for the National Academy of Sciences' 1975 COMRATE Committee report, "Mineral Resources and the Environment," illustrates some of the reasonable possibilities for energy growth, per capita consumption and population in the United States for 2000 A.D. It can be seen that the projections are from 69.1Q (slightly less than the 1972 figure) to 194Q

112

TABLE 3   Division of energy use in the 1972 U.S. reference system
($Q = 10^{15}$ Btu)

Prime Resource Component	Q	% of Total	Energy to Electricity			Energy to Consumer	
			Q	% Electric Total		Q	% Energy Supplied
Hydroelectricity	2.9	4	2.9	13.1			
Nuclear Energy	0.6	1	0.6	2.7			
Coal	12.6	17	7.8	35.5		4.6	8
Natural Gas	23.1	32	7.3	33.1		16	29
Crude Oil	33	46	3.4	15.6		29.6	53
TOTAL	72.2	100	22	100		50.2	90
Total Electricity, 6.3 Q less transmission loss (~10%)						5.7	10
TOTAL						55.9	100

Notes

(a). The overall efficiency of the electrical system, neglecting energies for prime source extraction and transportation is given by the ratio 22 Q in to **5.7 Q** out, i.e., 26%. The Overall efficiency of the Light Water Reactor, including all components except end use is 16.3% (60).

(b). Coal accounts for 35.5% of prime source energy for electricity and the electrical industry uses 62% of the country's coal supply.

(c). The electrical industry consumes 30% of the nation's prime energy and the largest share of its energy capital to supply 10% of the total energy supplied to the consumer.

(d). Hydroelectricity uses a continuous prime source and a fraction of this supply represents storage. This storage component is a conservation item.

TABLE 4   Components of the 2000 AD U.S. energy system and their potential
contribution to the total energy supply if all developments were acceptable
and fully exploited.*

Type	Take Off Date	2000 AD Resource Consumed
Nuclear LWR		16.5
Hydro		3.7
Coal		44
Oil (Steam Electric)		1.9
Oil (Domestic and Imports)		19.7
Natural Gas (Steam Electric)		--
Natural Gas (Domestic and Imports)		22.8
Solar**		
Photovoltaic	1990	23
Ocean Thermal	1995	6
Solar Thermal	2000	4
Biomass		15
Solar at Consumer		
Heating and Cooling	1990	12
Wind	1990	17
LMFBR	1993	3.9
GCFBR		--
HTGR	1980	3.9
Geothermal		6.6
Fusion		--
Oil Shale		8
Waste		6.5
TOTAL		214.5

*   Based primarily on ERDA 48, Scenario 5.
**   Project Independence Accelerated Development Projection.

TABLE 5   Goals of the U.S. National Program of Energy Research, Development and Demonstration and the possible impact on the 2000 AD energy system.

1. Expand Supply   Impact in 2000 AD (Q)

   Enhanced Recovery - Oil and Gas   13.6
   Above Expected   Oil Shale   7.3
   2000 AD Supply   Geothermal (average)   4.2

2. Develop Inexhaustible Resources

   Solar, Fusion and Fission Breeder   ?

3. Transform Fuels to More Desirable Forms

   Coal - Direct Utilization Utilities/Industry   24.5
   Waste Materials to Energy   4.9
   Gaseous and Liquid Fuels from Coal   14.9
   Fuels from Biomass   1.4

                    TOTAL SUPPLY ENHANCEMENT   70 Q

4. Improve Efficiency and Reliability of
   Energy Converters and Delivery Systems

   Nuclear   28
   Steam Electric (non-nuclear)   2.6
   Electric Transmission   1.4
   Hydrogen Economy   ---

5. Improve End Use

   Solar Heating and Cooling   12
   Waste Heat Utilization   4.9
   Electric Transport   1.3

6. Increase End Use Efficiency*

   Transportation   9
   Industry   8
   Buildings and Consumer Products   7.1

       TOTAL SAVINGS THROUGH CONSERVATION STRATEGIES   74 Q

* Fuel cells should also contribute according to industrial sources.

115

TABLE 6   Projected gross energy consumption in the year 2000 (in quadrillions of Btu) for the U.S. under selected assumptions concerning population and per capita consumption.

Assumed Per Capita Consumption (10^9 cal. or 10^6 Btu)	Projected Population*	(Millions of Persons)	
	252	265	285
69.1 or 274.4 (1965 value)	69.1	72.7	78.2
87.8 or 429.1 (1970 value)	82.9	87.8	93.8
171 or 680** (Implicit in JCAE)	171	180	194
154 or 610 (JCAE with 10% conservation)	154	162	174

Total 1970 Gross Consumption was 17.3 x 10^{18} -cal. (68.8 x 10^{15} Btu)

*   The population projection of 252 million assumes a constant net reproduction rate of 0.9 from 1970 to 2000, the 265 million figure assumes a constant NRR of 1 from 1970 to 2000, and the 285 million figure assumes a 15% increase in age specific fertility 1970-1985, after which age specific rates are assumed to be consonant with an NRR of 1. All the estimates include a constant 1970 level of migration.

** The per capita consumption of 171 x 10^9 cal. (680 x 10^6 Btu) was derived from the Dixy Lee Ray projections of 49 x 10^{18} cal. (194 x 10^{15} Btu) gross consumption and the JCAE population of 285 million.

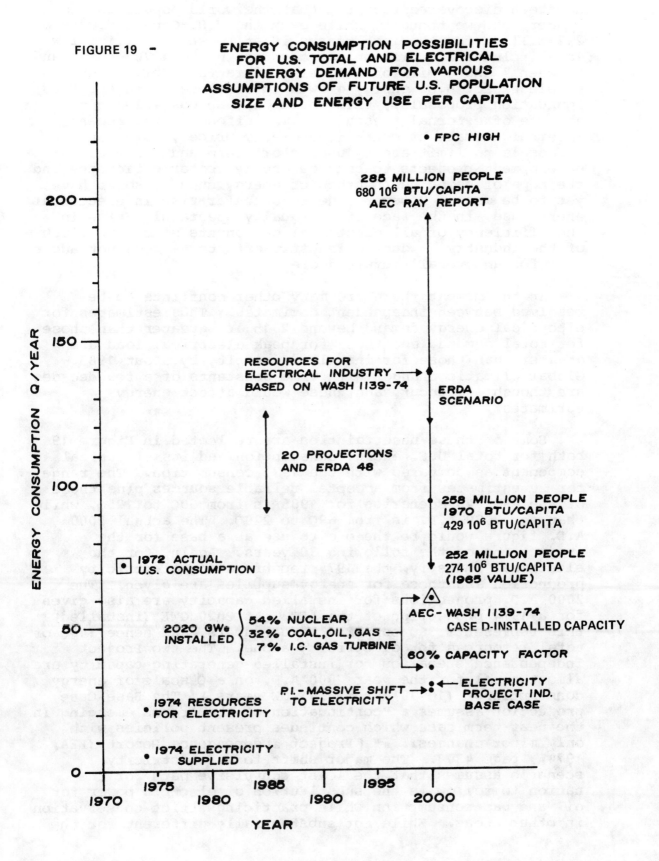

FIGURE 19 — ENERGY CONSUMPTION POSSIBILITIES FOR U.S. TOTAL AND ELECTRICAL ENERGY DEMAND FOR VARIOUS ASSUMPTIONS OF FUTURE U.S. POPULATION SIZE AND ENERGY USE PER CAPITA

117

for populations of between 252 and 285 people. In fact, it has been discovered recently that the world population is bigger than we thought, while even the U.S. Census figure of 212 million for the last census is at least 12 million low. This includes 5.3 million not counted in the 1970 census and between 7 and 8 million illegal immigrants. While most people expect U.S. energy growth to continue (even the Ford Foundation proposals), this would only be possible in the absence of external pressures. The effect of the recent internal climate of controlled energy prices, taxation proposals to limit energy use, short-term and long-term development projects to enhance energy sector efficiency and the hope of using new sources of energy such as solar have yet to be seen. However, the spectacular rise in electrical energy use, in the face of an equally spectacular rise in the efficiency of all electrical components since the birth of the industry -- due to industry efforts -- does not augur well for an overall future decrease.

At the moment there are many other conflicts to be resolved between independent estimates. Thus estimates for electrical energy demand beyond 2015 are greater than those for total demand, and those for peak electrical load are greater than those for installed capacity by about 1985. Global climatic changes with time constants of a few decades are thought to occur, and these would affect energy estimates.

Some of these uncertainties are reflected in Figure 19 both for total U.S. energy consumption and its electrical component as compared with the 1972 consumption. The range for 20 estimates from accepted reliable sources plus those of the ERDA 48 scenarios for 1985 is from 90Q to 125Q, while that for 2000 A.D. is from 69Q to 220Q. The actual 2000 A.D. figure would be the one to use as a base for the projections of the following 30 years. Again, for the electrical industry, the 1974 resources for electricity production and those for energy supplies are given. The 2000 A.D. projections for installed capacity are also given for a 1974 projection of the AEC. At 2020 GWE (installed) this represents 36Q supplied to consumers, and hence 144Q of total resources for electricity alone. The two Project Independence scenarios for installed generating capacity are almost equal for the year 2000 A.D. on a Q scale of energy consumption. ($1Q = 10^{15}$Btu$=33.43$ GW years.) The Base Case projection assumes a "continuation of the trends emerging in the near-term case which continues present policies with only minor changes...." [Project Independence Report (FEA, 1974), page 430]. The major shift to an electricity scenario assumes that the least convulsive path for the nation to follow is the substitution of electric power for oil and gas consumption while practicing strict conservation in other areas. While not substantially different for the year 2000, the difference between the two cases for the year 2030 is that between 2,000 GWE (Base Case) and 3,000 GWE (Massive Shift), i.e., 1,000 GW or 30Q, representing the difference between growth rates of 1.6 percent and 2.5 percent per year. The validity of these assumptions might be questioned if we remember that the historical growth rate in generating capacity has been about 7 percent per year. Whether or not such an abrupt change can be effected is open to doubt.

Assuming 1,500 GWE of installed capacity for the year 2000 A.D. as an average for both cases, this represents 900 GWE continuous supply at 60 percent capacity factor, corresponding to 27Q of electricity supplied to consumers and 104Q of basic energy resource required for electricity. The uncertainties are obvious. Thus, as an example only, if 104Q were needed for the electrical industry, and the country were to use 137Q as in preferred ERDA scenario V, (ERDA, 1975b) the consumer would get 27Q of electricity and 33Q from other energy sources, for a total of 60Q.

Some Global Uncertainties

World population growth is shown in Figure 20. In fact, world population is growing faster than is shown in Figure 20, since it reached 4 billion in early 1974. It is hard to believe that it can be turned down or levelled off at this late stage without catastrophe. Thus world pressures for resources may reduce our options severely, and we may have to rely on every technological option and domestic resource we can develop if we want to maximize our chances for survival as a society. A final uncertainty is the threat of nuclear catastrophe. This is the MAD world where the acroynm stands for Mutually Assured Destruction, the policy of the two superpowers.

Some Resolutions on Helium Conservation

Returning once more to the specific helium case which-- as I have tried to show--is part of this entire package of the U.S. and world energy and policy dilemma, it should be remembered that many prestigious bodies have advocated independently that our helium resource should be conserved. It is strange that a government which purports to be of the people, for the people, and by the people should ignore this call. The list of those advocating additional storage is long; it includes the U.S. Senate and House, the President's Energy R & D Advisory Council Resolution (1974), the 1969 Study and Recommendations of the Committee on Resources and Man of the National Academy of Sciences, the International Union of Pure and Applied Physics, the Academy report on

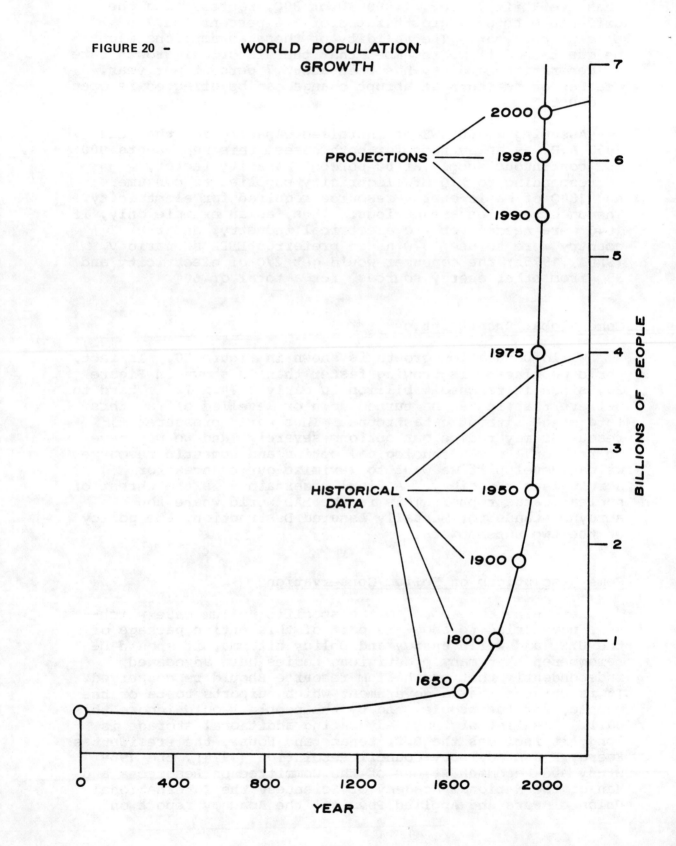

FIGURE 20 –   WORLD POPULATION
                          GROWTH

120

Physics in Perspective   (Physics Survey Committee, 1972),
the American Physical Society, the American Chemical
Society, the Argonne National Laboratory NSF Study on Helium
Use and Storage, and the Energy Research and Development
Administration's 1975 Study on the Energy Related
Applications of Helium.

Conclusion

In conclusion, it only remains to be stated that it
should be apparent that the developments in helium-based
technology are no longer so speculative as was thought by
those outside the field some years ago. We have been warned
that the crisis humanity will face during the next decades
will be by far the severest encountered during its two
million years of history, and that the present rate of
growth in population as well as energy and resource
expenditure cannot be kept up beyond the turn of the
century. If we believe that we should accept some
responsibility for trying to assure the future survival of
our children and our childrens' children; in other words, to
recognize and guarantee our future immortality through the
continued future existence of our society we need all our
options. At a cost of some $50 million per year in a
country with a GNP in excess of one trillion dollars,
guaranteeing the options opened by conserving the helium we
have paid for and managing the helium program intelligently
seems a small step to take. If we have no such belief; there
is no helium problem because most of today's leaders will
not be alive 30 years from now.

References

Physics Survey Committee. 1972. Physics in perspective, Vol. I.   D.A. Brom-
    ley [Chmn.]. National Research Council, National Academy of Sciences.
    Washington, D.C.

U.S. Energy Research and Development Administration. 1975b. A national plan
    for energy research, development and demonstration: Creating energy
    choices for the future. June, ERDA 48, Vol. I.

U.S. Federal Energy Administration. 1974. Project Independence Report,
    Washington, D.C., p. 430.

DISCUSSION

FRANCIS: I have a question on the Energy Doubler at Fermi Laboratory. What was the original cost of the accelerator compared to the present cost of adding the superconducting ring, then the cost for energy to run the accelerator with normal magnets and the projected cost to run the accelerator with the superconducting magnets?

LAVERICK: The Fermi Laboratory was built ahead of schedule, under appropriation and with more capability than in the design proposal. Original plans called for the accelerator to be completed and operating by June 30, 1972. Actual operation at 200 GeV was achieved in March, 1972. Full operation of experimental areas occurred in 1973, while the design report called for completion by 1975. The Congressional appropriation for Fermi Laboratory was $250 million. The final construction cost of $243.5 million provided an accelerator with more than twice the maximum energy specified, and also provided substantially more facilities than originally planned. For example, there are now four experimental areas in operation rather than the three of the original design proposal. Furthermore, a 15-foot bubble chamber, the largest in the world, originally estimated to cost $30 million beyond the appropriation for the construction of the laboratory, was included within the basic construction project. The super-ring was funded as an R&D project at a cost of about $40 million. It would take about an additional $12.8 million to convert this super-ring into the full Tevatron facility and a further $10 million for highest priority projects to exploit the Tevatron for a fixed target, TeV Physics.

The electrical energy cost for operating the machine depends upon the mode in which it is operated and the final energy of the protons. Since this machine has operated in various ways throughout the year, it is not possible for me to give a precise answer to this part of the question. However, in the energy-saver mode of operation, the new accelerator system could be operating much as it is in the present, but as a result of the loss-less magnets, the cost of the electricity for the laboratory could be reduced by $15 million a year. This is possible because the conventional magnet ring can be operated at a much lower level, and the protons injected into the super-ring which is practically loss-less for accleration to higher levels.

BIRMINGHAM: Mr. Laverick has given military applications, and he has pointed to a number of the civilian applications that are coming onstream in the electric power industry. One aspect that has not been touched on is how to assess the export market. You mentioned the Japanese. This country has a long way to go to obsolete the wheel, but the Japanese

122

have nearly done it: they are just completing a 7 kilometer test track for their high-speed train which is already running now at 200 or 300 kilometers per hour. They hope to go to 500 or 600 kilometers per hour by this time next year. They might need reasonable quantities of helium for the superconducting magnet system that levitates these vehicles, if they decide to go forward with it. The demonstration has already been conducted. I think it is only a matter of engineering and the economics of whether such a thing is really in Japan's national interest. This could result in a fairly large export market.

DRAKE: You didn't mention the date that they expect to finish their whole 7 kilometers of track. It is two years, isn't it?

BIRMINGHAM: As of last June or July the Japanese had completed half of it. They expect to complete the entire 7 kilometer test track by the end of this year. They already have a full-size and half-length car that they have run 200 kilometers per hour. On the longer test track the car can actually make up to 500 or 600 kilometers per hour. This track is similar to the Federal Railway Administration's test facility at Pueblo, Colorado.

DRAKE: What happened to their 1981 date for having a train?

BIRMINGHAM: I think that probably depends on the outcome of these tests and whether Japan then launches a national program to go full speed into development.

My point is that there are many developments that one needs to look at. If even two or three of them go forward, all the helium we have in reserve may be needed.

LAVERICK: Although I have looked mainly at U.S. projects, some fancy things are being done in the Soviet Union. I assume that if we were in a pinch we wouldn't give the Japanese or the Germans any helium for their trains and the Japanese would probably have to go elsewhere for it, probably to the Middle East. I think they have already opened some negotiations there. There is the possibility, however, that if you want to get rid of United States helium, you can find a market for it.

DERBY: Is Japan now operating rotating equipment such as superconducting rotating equipment on a routine basis in power plants?

BIRMINGHAM: Within the last month I got a paper on Fuji Electric. They are doing work in what we would call central station generating plants, and I don't think that they are as advanced as GE, Westinghouse, or our own Electric Power

Research Institute, but they are coming up to speed fairly rapidly. This paper was describing a prototype development of 6.25 megawatts, I believe, and what the United States has is around 5 and going up to 20 megawatts.

The Navy is looking at shipboard installations, as Mr. Laverick mentioned, and one installation on the _Enterprise_ takes 4 gigawatts. That is twice the power consumed in Colorado, and so one can only speculate as to where this might go. I think the potential for superconducting motors and generators for mobile shipboard propulsion is enormous.

An Overview:
MANAGEMENT OF HELIUM RESOURCES

Thomas A. Henrie
Associate Director
Mineral and Materials Research and Development
Bureau of Mines

In his book Helium, Child of the Sun, Clifford Seibel,
who for many years managed the government's helium program,
related the story of how he presented a paper entitled "The
Rare Gases of Natural Gas" before the American Chemical
Society at its annual meeting being held in Kansas City,
Missouri, April 1917. When he had finished reading the
paper, he expressed regret that the study had no practical
application. Richard B. Moore, then superintendent of the
Bureau of Mines station at Golden, Colorado, suggested that
a "practical application" was at hand -- helium for
airships! At that time, the idea sounded ridiculous to Dr.
Seibel, for he had in his possession practically all of the
helium available in the United States and had sold only
small amounts for experimental purposes at the rate of
$2,500 per cubic foot. It would have cost more than $200
million to fill a single dirigible! Today, helium sells for
$35 per thousand cubic feet when purchased from the
government and for about $25 per thousand cubic feet when
purchased from private industry. Many dirigibles and blimps
have been filled with helium at about this cost. But it is
a fact that eventually, at some point in time, be it 50,
100, or 500 years from now, we will of necessity have to
extract helium from the air; absent a technological
breakthrough, this cost has been estimated at $1,000 to
$3,000 per thousand cubic feet. Will we be able to afford
the practical uses of helium at this increased price?

Helium is an inert gas that is essentially devoid of
chemical properties. It is the second lightest of the
elements. Helium, in particular, has some physical
properties for which there are no parallels. Of particular
interest is the boiling point, $4.216°K$. These unique
properties make helium useful for shielded-arc welding, for
breathing mixtures, for deep-sea diving and medical
purposes, for pressurizing and purging in the space program,
for controlled atmospheres for growing transistor crystals,

as a heat transfer medium for nuclear power generators, as a lifting gas, in leak detection, in chromatography, and in the developing field of superconductivity.

The element was discovered spectroscopically in the chromosphere of the sun in 1868 by Lockyer, who therefore called it "helium," from the Greek word for the sun, "helios." Helium is produced by radioactive decay of certain elements. Decay of radioactive elements continues to take place in nature, and thus there is a continuous production of helium. However, helium has only been concentrated in certain geological formations, and its availability is somewhat limited.

At the present time, natural gas--primarily fuel gas--is the only known economic source of helium. Although the atmosphere contains an inexhaustible supply of helium, its recovery from that source in large quantities is, with present technology, not attractive because of excessive cost and energy requirements. Our national helium resource base is, therefore, a minor constituent of the finite natural gas resource base. This has created an unusual situation wherein the helium resource base is dissipated without regard to the demand for helium.

In 1960, after extensive study by the Department of the Interior and other federal agencies, legislation was submitted to the Congress to authorize the Department to conserve helium for future federal government use through long-term (25-year) helium purchase contracts. Analyses of then known helium resources and expected future helium demand indicated that after about 1985, helium available from remaining known resources would not be adequate to meet the demand expected to materialize in the future. The idea was to recover some of the wasting helium and store it underground in an existing natural gas reservoir near Amarillo, Texas, where the government had acquired exclusive rights to the helium-bearing natural gas in the Cliffside field.

Subsequently, the Helium Act of 1960 was approved. The Act authorized the Secretary of the Interior to continue the operation of helium production plants and additionally to enter into up to 25-year contracts for the purchase of helium for conservation. Included within the Act was a section that stated: "It is the sense of the Congress that it is in the national interest to foster and encourage individual enterprise in the development and distribution of supplies of helium, and at the same time, provide, within economic limits, through the administration of this Act, a sustained supply of helium which, together with supplies available or expected to become available otherwise, will be sufficient to provide the essential Government activities."

All costs of the program were to be paid out of the income derived from the sale of helium. During the early years of the program, it was anticipated that expenses would exceed income from sales. Borrowing from the Treasury was authorized on an annual basis to provide the funds for helium purchases which could not be paid for from income.

It was expected that income would exceed expenditures by 1971 and that repayment of the funds borrowed from the Treasury would begin at that time. The Act required that all borrowed funds and the government's investment in capitalized property must be repaid, together with accrued interest, within 25 years. The Secretary of the Interior, under the Helium Act, has the option of extending the payout period by 10 years.

In August 1961, annual contracting authority of $47.5 million was approved by the Congress for the purchase of helium. By November 1961, four fixed-price, 22-year, take-or-pay contracts were negotiated and approved.

At the time the contracts were made, it was estimated that the government would buy a total of 62.5 Bcf of helium. It was expected that the Bureau of Mines plants would produce 15 billion cubic feet during the contract period. Helium demand was expected to be 36 Bcf, leaving 41.5 billion cubic feet of helium for use after about 1985 for essential government activities. Currently, the volume in storage is 39 Bcf.

In November 1961, the price of helium from the Bureau of Mines was raised to $35 per thousand cubic feet for all customers. Previously, federal agencies were charged $15.50 per thousand cubic feet and non-federal customers were charged $19 per thousand cubic feet. The increased price was calculated to liquidate all costs of the program during the life of the helium purchase contracts. Concurrent with the price increase, a single private plant began helium production. Shortly thereafter, additional private helium plants were built, and by 1966, there were seven. The average selling price from the private plants was about $25 per thousand cubic feet--$10 less than the government price. This private production and the disparity in price resulted in a rapid erosion of government sales, and loss of revenue to finance the helium conservation program.

An even more important factor was a decline in government usage. Helium demand had developed about as expected until 1967. However, in 1967 federal helium demand began to decline primarily because of slowdown of the space program and general decline of helium use by the government. This decline in demand, together with discovery of substantial new helium resources and improved helium extraction technology, led the Secretary of the Interior to terminate the helium purchase contracts in 1973.

However, for various reasons, four of the five conservation plants continued to produce "conservation helium," some of which was stored, while the remainder was either vented or returned to the natural gas going to the fuel market. Several reports and congressional resolutions have addressed the problem of saving some or all of this helium being vented to the atmosphere for future beneficial use.

The Bureau's responsibility, under the law, is to provide a sustained supply of helium to meet foreseeable needs of federal agencies. The helium in storage and reserves known to exist on the federal lands will meet federal demand well into the twenty-first century. There is no justification, and indeed no authority, for further purchases of helium by the Bureau, since such purchases would be aimed primarily at meeting future private sector needs. The basic helium problem, therefore, is what actions should be taken to meet future helium requirements in the private sector?

The Bureau of Mines took what action it could under current laws to encourage conservation of helium by private industry. In 1975, the Bureau revised its policy concerning storage of private helium by substantially reducing storage charges and deferring payment of the bulk of the charges until the stored helium is withdrawn for sale. Since revising its policy, the Bureau has entered into 10 new storage contracts with private firms and has accepted a total of 1.2 Bcf of helium for storage under these contracts. We are convinced that the Bureau's charges are not a significant factor deterring storage. The storage costs for 20 years will amount to only 4 percent of today's sale price of helium.

However, storage of helium for future use involves a heavy financial risk because of litigation concerning the value of helium at the wellhead. Particular attention must be directed to the lawsuit Ashland Oil, Inc. v. Phillips Petroleum Company. The purpose of the lawsuit is to determine the reasonable value of helium contained in natural gas sold by Ashland to Phillips. The court found that the reasonable value of the helium prior to extraction and sale to the United States ranged from about $12 to $17 per thousand cubic feet. The Tenth Circuit Court of Appeals, on May 11, 1977, affirmed the method by which the District Court arrived at a value of $12 to $17 per thousand cubic feet of helium at the wellhead. Under the District Court's method, most of the revenue from future sales of stored helium would accrue to natural gas producers and

128

owners of the land containing deposits of helium-bearing natural gas rather than to the companies which extract and store the helium. It is doubtful that private companies could afford to save helium for future use in the face of such an allocation of the sale price.

The decision in this case is also important to the United States, because the contract for the sale of helium to the United States by Phillips, as well as contracts with other companies, contains a provision which may require the United States to indemnify Phillips for payments to third parties for helium value in excess of about $3 per thousand cubic feet. If the value of $12 to $17 per thousand cubic feet as determined in this case is finally upheld, it may result in claims against the United States amounting to $9 to $14 per thousand cubic feet in addition to the average of about $12 per thousand cubic feet the United States has already paid for the helium.

As stated previously, the helium problem is basically a problem of supply and demand in the private sector. To fully understand this situation, we must look at the resource base and predicted future demands.

The helium resource base for the nation is calculated by the Bureau of Mines by combining the American Gas Association estimate of proven natural gas reserves, and the Potential Gas Committee estimate of the potential natural gas supply with the Bureau's knowledge of natural gas helium contents. The Bureau has been evaluating the helium resource base since 1920.

Presently total identified and undiscovered helium resources in the United States are estimated at 704 Bcf. Of this volume, 186 Bcf is classified as reserves, 117 billion cubic feet of which is known to be nondepleting. The remainder, 587 Bcf, which includes future discoveries, is assumed to be in the depleting category. Of the 117 Bcf which is nondepleting, 45 Bcf is dedicated to the Bureau of Mines and stored. And of the other 72 Bcf, about 60 Bcf is located on federal land and subject to P.L.-145 which reserves ownership of all helium underlying federal lands to the federal government.

It is estimated that about 140 Bcf of helium could be recovered from natural gas produced in the United States between now and the year 2000, in addition to about 34 Bcf which is recoverable at existing helium plants.

Periodically, the Bureau also surveys the nation's helium users to obtain future demand forecasts. The most recent forecasts indicate that the total U.S. demand for helium in the year 2000 will be double the current demand.

Of the 117 Bcf of helium which is nondepleting, as stated previously, approximately 39 Bcf is in the Cliffside storage field. To purchase this helium, the Secretary of the Interior had to borrow $250 million. Interest over the intervening years has increased the debt to approximately $470 million. Final judgment in the helium title lawsuits could result in additional claims against the government amounting to several hundred million dollars.

In conjunction with its determination of the nation's helium resource base, the Bureau has for some time been looking at the helium resources outside the United States. Although this information is limited, known helium resources of the world excluding the United States and countries with centrally controlled economics are estimated to be 184 Bcf, of which 89 Bcf is in Algeria, 39 Bcf in Canada, 23 Bcf in Australia, and 12 Bcf in various other countries. To assist in compiling the foreign helium resource data, the Bureau of Mines has collected 382 gas samples from 36 foreign countries.

What is the helium problem? It is a problem of supply and demand without parallel--a relatively low present-day demand and dissipating known resources versus possible high demand and undiscovered resources in the future. It is a problem of either storage now for future use--with inherent economic and judicial factors involved--or of not storing now and relying on present-day nondepleting resources, future discoveries, and eventually the atmosphere.

DISCUSSION

LONG: One would like to think that this roughly 40 Bcf of helium now in the Bureau of Mines storage facility might be kept as reserve for a relatively long time. I'd be interested in your personal judgment as to whether it would make sense to hold that amount in full reserve for the next 50 to 75 years? Obviously coupled with that question is: What does the law say, and what changes in the law might be necessary if we decide to do that?

HENRIE: At the present we only have authority to maintain necessary government requirements for helium, and we have the responsibility to sell helium. We have a contract that we must fulfill, in which we are extracting helium from natural gas and selling high-purity helium. We also generate some helium from the Cliffside field in maintaining the field. But under the present law, we have no authority and no funds   to protect the requirements of the federal government except that reservoir and these other supplies. I think Mr. Tully can best answer your question. Some statistics that ought to be brought out concern the present flow of helium. We will limit this to that area in which we have some means of taking care of the flow. That would be those production facilities around the pipeline. I think it is very difficult to talk about any helium being wasted or vented that is not in some way connected to the pipeline, because you could not extract it and take it over and run it through the process. If Mr. Tully would come up and go through the present flow of helium on a yearly basis, I think these numbers would best answer your question.

PHILIP TULLY, U.S. Bureau of Mines: These statistics will briefly summarize the entire flow of helium in the United States on an annual basis. For the Bureau of Mines, we're selling approximately 200 million cubic feet to federal agencies and we are storing about 150 million cubic feet. All of this production is from our Keyes plant. The private crude helium extraction plants are performing approximately as follows.

Northern Helex Company is recovering about 600 million cubic feet per year; they are storing about 400 million of that, or two-thirds. A portion of the balance is being stored and a large portion of it, about 160 million cubic feet, is being purified.

Phillips Petroleum is presently venting about 0.3 Bcf but, as you heard this morning, they intend to commence storage of that. The balance of their production goes to Alamo Chemical Company, a subsidiary of Phillips.

Cities Service Helex, in conjunction with Cities Service Cryogenics, is extracting about 700 million cubic feet a year, venting about six-sevenths of that amount; the balance, the one-seventh, is going to Cities Services Cryogenics for purification, liquefaction, and sale. National Helium Corporation and the Phillips Dumas plant are not presently extracting helium.

We also have private plants for producing pure helium. Cities Service Cryogenics is extracting about 150 million cubic feet of helium directly from natural gas and taking about 75 million cubic feet from Cities Service Helex Company. The sum total of that, about 225 million annually, is marketed in the liquid form for distribution purposes. Alamo Chemical-Gardner Cryogenics is extracting approximately 140 million cubic feet from natural gas; Kansas Refined Helium Company about 30 million; and Western Helium Corp. in New Mexico, in the Penta Dome-Four Corners area, about 30 million cubic feet. To recapitulate, the entire pure production and sales is approximately 850 million cubic feet, of which 170 million is exported, and about 680 million cubic feet is consumed annually. About 0.6 Bcf is being stored presently by Kansas Refined Helium

Company and Northern Helex, and about 1 Bcf is being vented. The total amount of helium thus recovered, utilized, and vented is just slightly under 2.5 Bcf.

EARL F. COOK, Texas A&M University: I think you said that it is estimated that 140 Bcf could be extracted between now and the year 2000. The Bureau's estimated helium reserves in helium-rich natural gas as of the close of 1973 were 119 Bcf. Those had been decreasing at a rate of about 8 Bcf a year. My question is: Where is the 140 Bcf going to come from? Does that include new discoveries of helium-rich natural gas?

TULLY: The 140 Bcf referred to comes from all qualities of natural gas. It is the ultimate volume estimated to be recoverable if we recovered all of the helium from all of the natural gas that could be economically recovered. I use economical in the sense that these supplies of natural gas come from existing gathering fields, and on and in pipeline systems to the extent that the gas is consolidated sufficiently well to warrant the establishment of new construction facilities if that were necessary.

COOK: I'm still not quite clear as to whether that is based on proved reserves of natural gas or whether it goes beyond proved reserves to include some of what the Potential Gas Committee likes to call potential reserves and the Bureau of Mines and the Geological Survey prefer to call resources.

TULLY: I'm not exactly clear as to whether that does encompass those numbers. I believe that the proved reserves published by the American Gas Association are approximately 200 to 220 trillion cubic feet, which, based on averages talked about this morning, would yield something slightly less than 200 Bcf of helium. It largely depends on where these gases are coming from, and I do not remember breakdown as to the respective percentages of the 140 Bcf that would come from proved reserves or probable, hypothetical, speculative, and as yet undiscovered natural gases.

LONG: If I may go back to my first question, I'd be interested in either your personal philosophy or the Bureau's philosophy about the concept of building a long-time reserve. Does it make sense to build a strategic reserve of helium that is not confined to a period so short as the current law but would be available up to 2025, or some such date?

HENRIE: I don't know the Bureau's long-term philosophy. I've only been involved with this seven years myself. I think within the Bureau -- and this is outside of any litigation or any considerations of this nature -- we believe that there ought to be some changes in legislation whereby this situation that we're in now could be removed and Congress could work out something so that helium could be stored. Oftentimes, as we've discussed this, we've felt that Congress ought to be the responsible body to determine how much should be stored after it has received the advice of the Secretary. This would put it more into a stockpile-type situation than we have under the present law. This is the kind of thing that is often discussed around the table: What do we do about the helium problem? That's about as simple as I can put it. We know that it's a serious problem; we look down the road and wish we had a simple answer. I don't think the present law is a simple answer. And I have found that feeling in the appropriations committees that I've been before. Things should have been different.

HULM: Addressing the question that Franklin Long is asking, it seems to me as we look at the economics of the helium storage program that, as you have said, the original plan was that the income would exceed expenditure by 1971. I believe, as I recall from the wording of the Act, that all of the indebtedness plus interest was to be paid back within 35 years. One can look at this as a kind of financial plan. It is the sort of plan that businesses make when they go into a long-term operation, especially into a new venture. They have to have a period where they're going to go into the red, as you did, and then at some point you've got to cross over into the black. I get the impression that the thing that derailed the program in the sense of stopping conservation in 1973 was the fact that the reappraisal of the financial plan indicated that it had gone wrong, for the reasons that you also stated -- namely, that the sales were a factor of two less than anticipated plus the fact that private suppliers had entered the market. And so the income had no hope of matching the outgo in 1971.

It seems to me that one should in those circumstances prepare a revised financial plan. Clearly the payout on this is much longer term than it was originally thought. Various studies of this problem have been made available-- one in particular by Dr. Howland, who talked this morning.

Do you have a financial plan that calls for payback at a later date, based on projections of demand for helium and some kind of strategy with respect to the price?

HENRIE: In about 1974 we went into great detail before the appropriations committees and told them about our plight; we spent two hours in testimony before both committees. Because of that testimony, we now have Congress taking a greater interest. I personally feel that this is the purpose of the request for this present report. Congress would like to get as much information as possible in regard to new legislation. My own personal feeling is that we're all quite frustrated with the prsent law -- not only the Secretary and his staff, but also the people on the Hill. I just don't believe that there's an easy solution. We've batted quite a few things around; we've been as responsive as we know how to the various committees that have asked us.

HULM: I feel a little frustrated with that answer. I think it would be helpful to this Committee if you could give us some idea as to what changes in legislation might be made.

HENRIE: Congress asked us to prepare a report by January 31 of 1978. In that report it's necessary for us to make recommendations. We felt that one source of information and a good area to obtain data, information, and philosophy would be from the National Academy of Sciences. We would hesitate to make recommendations to you to make recommendations back to us. We would like to brief you and fill you in with as much information as we can, but we'd sooner not bias you, because it'll be necessary for us to comment on your recommendations.

WILLIAM D. CAREY, American Association for the Advancement of Science: It might be helpful if the Academy knew your state of mind so that we could advise you about your state of mind as well as about our own. Do I understand that the Department of the Interior, if we can scale it up from the Bureau of Mines for a moment, is unable to make a policy proposal in this area because of the litigation? Is that a handicap to any macropolicy-making?

HENRIE: We'll have to clear our report through the Secretary's office, which means it will have to go through the Solicitor's office. We also hope to do this with DOD, NASA, and DOE. But I'm sure that Congress would like our recommendations to be outside of the area of litigation, and I feel certain that the reason they've asked for the report is to gather information to write a new law. There is a lot of interest in trying to solve this problem. We would be most happy to bring staff in from Amarillo to sit down with this Committee and go through an orderly briefing in detail. The Government InterAgency Committee went to Amarillo for a briefing and learned a great deal.

CAREY: If I thought that wisdom would envelope this Committee by going down to Amarillo, I think I'd recommend that we be on a plane tonight. I'm not sure it's that easy. Many years ago I used to be on the Bureau of the Budget. When I think of the limited time that this Committee has to embrace the issues before it, I am reminded of the rule that we quoted to the White House from time to time when they wanted an answer before lunch: "When you want it bad, you'll get it bad." I hope that doesn't apply to us.

All of the background papers say it's inevitable that eventually the atmosphere will have to be looked to as the source for helium. If that is indeed the case, what I'd like to know is what the Bureau of Mines or anybody else in government is doing to fund advanced research and development looking toward more efficient recovery methods for the helium now vented into the atmosphere, and what the prognosis appears to be for any breakthroughs in that direction, even though it may be a decade or so before they materialize.

HENRIE: Surely there's some relationship between concentration and cost. I'm not so sure that it's a direct relationship as was implied, which would indicate that some research should be done. In the case of taking helium from natural gas, the working fluid is methane, which works at quite a different temperature than if you took helium from air, in which case the main working fluid would be nitrogen. So we might be on a different curve than the direct implication from the direct cost. We know a lot about the cryogenic systems. I think the Bureau has done a lot to develop the cryogenic systems, materials to be used, transfer data, and things of this nature. We've used helium to help define equations of state and this type of research. I've never had brought to me the problem of would we use something other than a cryogenic system to separate helium from the other components. I'm sure that this ought to be looked at earlier than the 20 years or so from now when we will be faced with the problem of economically separating helium. But I do know that 5 ppm is not very much. You're handling a large amount of material. It seems to me that the economics of the industrial gases would have a lot to do with the production of helium.

Our research in helium within the Bureau is quite limited now compared to a few years ago. My own appraisal of the staff group involved in this activity, as well as that of two directors of the Bureau of Mines, is that we have a tremendous technical competence. We are quite proud of this competence. I think we should start applying some research funds to this program.

SMITH: I wonder whether we could learn anything concerning the prospects for technical change in extraction of helium from what has happened with other elements and their extraction processes. I'm neither a scientist nor an economist. I don't know anything about the prospects here. Have the relative costs of extracting liquid nitrogen or neon or other elements changed dramatically within the past 15 years?

HENRIE: Availability of many of these elements in liquid form is quite common now--more so than a few years ago. I don't know what the cost of argon is now, but I know that it is not a problem. Argon is obtained from the atmosphere by liquefaction processes. For those people building titanium plants and doing cost analyses in the mid-fifties, argon was available.

FRANCIS: A considerable amount can be learned from the experience of our industry in the extraction of helium from the atmosphere. Linde Division of Union Carbide operates plants today that extract neon-helium mixtures from the atmosphere; it also does air reduction. I think Mr. Gilardi might comment on that. A number of other industrial gas companies do the same. This is a by-product of our manufacture of oxygen, nitrogen, and argon. We are able to produce neon--present at 18 ppm in the atmosphere--together with about 22 percent helium along with the neon, and it currently is being utilized by the national laboratories in neon spark chambers, in the big neon-filled bubble chamber at Argonne National Laboratory, and so on. The price of neon for such uses in this crude neon-helium mixture form reached a level of about $.20 a cubic foot and was produced in substantial quantities--several hundred thousand cubic feet. I have written one paper on extracting helium from the air, and other papers were written by people from the other industrial gas companies. It's rather clear that insofar as we consider by-product helium from the atmosphere, the technology already exists, and in the time period around the turn of the century, if the price level goes to the point that we expect it to, a certain quantity of several hundred million cubic feet a year would probably come from this type of activity. But I think that the crux of the matter is not the question of by-products of the air separation industry, because there you're limited to the quantity of oxygen and nitrogen that is demanded by society as the prime products. This puts that limit at maybe a couple of hundred million cubic feet. That's the expected industry that will develop by the turn of the century, assuming a growth rate such as we have experienced in the past years. That future industry would be many times the size of the present air-separation industry. The crucial question becomes: How expensive is it to extract helium from the atmosphere as the primary product when the demand for helium sets the production rate? It is possible, because we understand the nature of the gases that we're dealing with, to sketch what an air-separation plant based on cryogenic technology would be like. This has been done and is part of the record--a plant designed for helium extraction as the prime product from the atmosphere. The crucial problem is the vast quantity of air that must be passed through that plant. We did consider the kind of a plant needed to do separation by known principles, and this resulted in the cost figure of $1,000 to $3,000 per 1,000 cubic feet.

I have given some thought to the question of what conceivable new technologies could be applied to this process so that the actual act of separation will not depend on cryogenic work but will depend on something else--perhaps laser stimulation of helium atoms so that they can be plucked out by a tweezer or something like that. The important thing is not how do we do it but is the concept of what do you get to when you have this black box. We haven't done the research. The question is: If we did the research, where would we get? The one irreducible cost is the cost of moving the atmosphere into the black box, through it, and discarding it out the other side. That cost alone--just moving the air through the black box--must be a minimum of 10 times the present extraction cost. I don't see how anybody can possibly get around it; you simply have to move huge quantities. I want to illustrate that by this comparison. The flow into the National Helium Corporation plant at Liberal, Kansas, was approximately a billion cubic feet of natural gas a day. From that they were able to extract, at the maximum flow rate, 1.3 Bcf a year. A plant to extract helium from the atmosphere as the prime product would flow 1.5 Bcf of air an hour into a plant that would produce 60 million cubic feet of helium a year. The flow of air through the separating equipment would be staggering and would be most of the cost. It is inconceivable to me that any technology can wipe out this cost.

LONG: If indeed the Department has in mind the possibility that a bill will result from this interagency report, I can only note that you must be--as we are--impressed by the complexities and the number of interested parties. I hope that the feelings of private industry, landholders, and prospective scientific and technical users get into this bill-making operation. It is a challenge but it is also a very interesting opportunity. In this process of looking at what to do, I'm sure that more attention will be paid in the future to this question of how much helium is available in the so-called nondepletable reserves. What is their solidity, their reality, and what might be the production costs from them? Is that sort of thing reasonably well understood within your group?

HENRIE: Yes, and I think that if you desire a briefing from our Amarillo people they could put this package together for you so that you could understand it very well.

ALBERT E. UTTON, University of New Mexico: This morning we heard from industry that one of the biggest uncertainties that they face is the fact that the government can control the price of helium in the future through this stockpile, this stored reserve. How might the government react and reduce that uncertainty in the future? Do you have any comment or reaction to that?

HENRIE: Certain people are advised on the stockpile, and someone else makes the decision. I know that it's a fear of many that the government buys, stockpiles, and then--when the prices start up--they dump the stockpile and use that as a means to keep prices down. That's why I think it's important that Congress decides and then is willing to appropriate the funds and determines the size the stockpile should be.

HULM: You laid stress on the fact that there's a great uncertainty in the area of payments or royalties to gas producers and landowners for the helium that has already been extracted. There are a number of suits in progress, I gather, on this issue; one of your fears is that because of the nature of the original contracts between the suppliers and the government that they contain a clause which makes the government liable for add-on royalties beyond $3.00 a thousand cubic feet. Isn't that correct?

HENRIE: As I understand now, litigation is before the Supreme Court. Is that not right? And how the Supreme Court rules in the government's case, I don't know. I understand however they rule that it's not our responsibility but the Department of Justice's responsibility to pay and acquire the funds.

HULM: I realize this is under litigation, and therefore it is difficult for you to comment. But it seems to me that it is one of the points brought up by one supplier this morning as to why they're not storing helium or taking advantage of the Bureau's storage policy. They feel great uncertainty regarding the ownership of that helium. It is like something out of Gulliver's Travels. If you throw helium away then you can't be responsible for paying anybody for it. That seems to be the philosophy. I'm not sure whether that's the correct understanding of the problem. But my question is really this: If there is any legislation written, would you expect it to contain anything relating to this question? Or would you prefer to see the outcome of the litigation before you answer my question?

HENRIE: We would sooner see people store it, to tell you the truth. We tried to make the storage costs reasonable. We wanted to encourage as much storage as possible. Some companies have chosen to store helium, and I suppose that each company has its own reasons to be one way or the other. I do not presume to know all the reasons why the situation is exactly the way it is.

HULM: Do you feel that you have done all that you possibly could to push them? Couldn't you give them a little more incentive to stop this wastage?

HENRIE: We have no funds to take the cost ourselves, and Congress is very willing to deny us an extension of the $47.5 million each year that goes through in a minute or two. It's a matter of what best can we do with the limited authority that we have, and we chose this system. I know the Director of the Bureau of Mines gave it a lot of thought and our people in Amarillo gave it a lot of thought before they put the package together. And we briefed the Director on it twice before he accepted it. As I recall, there was no hold-up in the Department at all to our recommendation. This was the best package that we could come up with in our own judgment.

UTTON: I wonder whether I could ask the same question again and maybe get a response from industry, maybe Mr. Culbertson, as to what they might recommend as a way of removing the uncertainty that they see caused by this storage of 39 Bcf of gas and the possibility that the government therefore may control future prices rather than the marketplace.

CULBERTSON: This morning I listed five impediments that we saw bothering people in private industry like ourselves in regard to storage of helium. The first impediment is the overhang of the large volume of helium already in storage controlled by the government and the potential for future price changes that we couldn't forecast. I don't mean that we expect price certainty; but actions by anyone who controls such a large percentage of the future supplies must be taken into consideration.

The second impediment is that expensing of the costs of extracting and transporting helium cannot occur until we sell the helium, as we understand the present IRS rules. The third one is the potential danger of ad valorem taxes levied by the states on privately owned stored helium. The fourth is the current litigation, which has had considerable discussion, and the fact that we're faced with uncertainty as to the payments due landowners and producers.

The last impediment is the term of a government storage contract. The only problem we have with that, Dr. Henrie, is whether it could be a longer term. Considering the future as we see it, it may well be more than 25 years before we can move the helium out of storage and this bothers us. We'd like to have a little longer shot at using the storage. That would help relieve some of our concerns.

The Bureau of Mines has been very cooperative and very helpful in trying to do what they can to encourage the storage of helium. Now as to what might be done from here on, we would like to see--and some of this we realize might not be politically acceptable--Congress act again to maximize conservation in helium, and it should again recognize, as it did in the Helium Act of 1960, that the recovery and long-term storage of helium can best be accomplished by a cooperative effort between government and private industry. We think the Congress should act swiftly to save and to encourage private industry to save as much as possible of the large volumes of readily recoverable helium now being dissipated. There are several steps we think should be taken that will result in increased helium. For example, Senate Bill 2109, recently introduced by Senator Pearson, in our judgment, will accomplish many of the steps that we think are needed if approved by Congress as it was introduced. One of the first things that needs to be done is removal of the five impediments listed earlier. Certainly this removal would encourage the long-term storage of helium by private industry, and we think most of these, probably not all, but most of these can be removed by legislation. As far as this $12 to $17 question about what would be paid to producers and landowners, I suspect that this is going to have to go through the court process. I don't really see any short way to solve that problem.

To conserve the present stores of government-owned helium now in Cliffside and to allow conservation of all the helium produced from government plants in the future, we would suggest that federal agencies obtain their major helium requirements from nonfederal sources. As I understand it, the current law requires federal agency major helium requirements to be supplied by the Bureau of Mines. We feel that the government should conserve its helium in storage until private industry can no longer reasonably supply current demands from depleting natural gas. This action would effectively conserve approximately 0.02 of a billion cubic feet a year of helium. And it seems to us this change could be accomplished within a reasonable amount of time.

We suggest as another step to be taken to further conserve helium that the government reinstitute the purchase of helium for storage. Plants that currently have excess producing capacity obviously would be immediate sources for such purchases. In addition, there are some helium-bearing natural gas streams flowing to markets from the Hugoton and Panhandle fields that have never been processed for helium extraction. Some of these streams contain helium in excess of 0.03 percent, and the helium extracted could be relatively inexpensive as compared to alternate future sources. I have seen a figure showing that to recover a billion feet of helium a year from the atmosphere would take half of the projected throughput of the Alaska pipeline. This is a factor not mentioned, but it would be tremendously expensive so far as energy requirements are concerned.

HUBBERT: I'm still somewhat concerned over the estimate of the future. Those figures of the Potential Gas Committee and the Geological Survey until recently were both running around 1,800 or 1,900 trillion cubic feet for the ultimate amount of gas from the lower 48 states, as I recall the figures. The recent Geological Survey Circular 725 by Miller et al. (1975),   cut that figure down to somewhere pretty close to 1,200, or about two-thirds. So if we subtract the past accumulated production--that from memory is somewhere in the neighborhood of 500 trillion--that leaves the difference between 700 trillion for the lower figure and about 1,300 trillion as the figure you're dealing with here, depending on whether you use the higher figures or the more recent ones. I strongly commend that the higher figures be abandoned, and that you base your estimates on those in the Miller et al. circular.

HENRIE: I respect your recommendation, and if this Committee asks to have us brief them further, we'll make sure that your question here is resolved and that they are briefed according to the background that you've asked for. As far as our Committee is concerned we'll probably put it in the report as the best information that the Bureau has at that time.

S. LOCKE BOGART, Department of Energy: My statement has been appropriately concurred upon by both the Divisions of Magnetic Fusion Energy and Laser Fusion Energy. Fusion reactors are expected to require substantial quantities of helium as a superconducting magnet and blanket coolant. Forecasts of the potential helium demand by fusion energy are provided in The Energy Related Applications of Helium (ERDA-13) on pages 37-39 and in Appendix E of that publication. In general, the analysis leading to the projected demands erred on the conservative side. The forecasted demand was substantial, reaching a cumulative maximum of 52 billion standard cubic feet by the year 2030.

Since ERDA-13 was published in the Spring of 1975, both the magnetic confinement and laser fusion programs have demonstrated further optimism for the prospects of this inexhaustible energy resource alternative. Physics and technology developments have occurred as expected, and there has been recognition of new energy applications that may broaden and accelerate the implementation of fusion energy.

However, neither of the two Department of Energy fusion programs have reached the point of making an unequivocable statement of the certainty of success. While there is a general atmosphere of optimism, the uncertainties of physics, technology, and economics compel the fusion communities to still "err on the conservative side." Many of these uncertainties are expected to be clarified within the next three to five years, if the required level of support is provided.

It is fortunate that the resolution of these uncertainties is expected in the relatively near term. If the current policy of venting helium to the atmosphere continues, then not too much will be wasted by the time the prospects for fusion energy are established. While we deplore this practice, we also cannot, in all conscience, request that helium be conserved on the basis of the expectations of success for fusion energy.

The fusion programs of the Department of Energy recommend the following:

1. Encourage the storage of helium at minimum cost to the involved interests. This will affect the federal government, the private sector, and state and local governments.

2. Periodically review the status of and prospects for advanced energy technologies that require substantial quantities of helium. This should be done at three to four year intervals.

3. Assess the resource base for helium, especially as it relates to its association with the size of the natural gas resource.

DRAKE: This concludes our Forum, and I want to say, on behalf of the Committee, that we are appreciative for the information and for the discussion. Now we will have to try to sort this out and come up with something meaningful. It is a formidable task made even more so by the short time available to us.

HELIUM FORUM
List of Attendees

Allen F. Agnew, Senior Staff Specialist, Congressional
Research Service, Library of Congress, Washington, DC
20540; (202) 426-5878

Earl K. Anderson, Grants and Contracts Specialist, National
Science Foundation, 1800 G Street, N.W., Washington, DC
20550; (202) 632-5884

Thomas Bicknell, Auditor, General Accounting Office, 7509
Buchanan Street, Apt. 118, Hyattsville, MD 20784; (301)
577-8798

Bascom W. Birmingham, Director, National Bureau of
Standards, Boulder Colorado Laboratories, Boulder, CO
80302; (303) 499-1000, Ext. 3237

Fred Block, Staff Officer, National Research Council, Board
on Mineral and Energy Resources, 2101 Constitution
Avenue, N.W., Washington, DC 20418; (202) 389-6508

S. Locke Bogart, Division of Magnetic Fusion Energy,
Department of Energy, MS G234, Washington, DC 20545;
(301) 353-5160

Mr. and Mrs. J.C. Boyce, Retired NAS Staff Member, 2500 Que
Street, N.W., Washington, DC 20007; (202) 337-8927

J.N. Brooks, Planning Consultant, 1401 PB Phillips Building,
Phillips Petroleum Company, Bartlesville, OK 74003;
(918) 667-6967

William D. Carey, American Association for the Advancement
of Science, 1776 Massachusetts Avenue, N.W., Washington,
DC 20036; (202) 467-4470

John A. Carr, Economist, Bureau of Domestic Commerce, Room
2007 - Main Commerce Building, Department of Commerce,
Washington, DC 20230; (202) 337-3758

C.A. Conoley, Attorney, P.O. Box 1348, Kansas City, MO
64111; (816) 753-5600

Earl F. Cook, College of Geosciences, Texas A&M University,
College Station, TX 77843; (713) 845-3651

Craig A. Coulter, Attorney, Cities Service Company, P.O. Box
300, Tulsa, OK 74102; (918) 586-2269

143

LeRoy Culbertson, Vice President, Phillips Petroleum
    Company, Bartlesville, OK 74003; (918) 661-4360

M.H. Cullender, Manager, Contracts Administration, Phillips
    Petroleum Company, 5 PB Phillips Building, Bartlesville,
    OK 74003; (918) 661-4357

J. Warden Cunningham, Director, Government Accounting,
    Airco, Inc., 1030 - 15th Street, N.W., Washington, DC
    20005; (202) 296-4775

Roger Derby, Program Manager, Division of Energy Storage
    Systems, Department of Energy, 600 E Street, N.W.,
    Washington, DC 20545; (202) 376-4745

Robert M. Drake, Jr., Studebaker-Worthington, Inc., 116 N.
    Upper Street, Lexington, KY 40507; (606) 253-0762

Dan Edwards, Chief, Ecologic-Economic Analysis, U.S. Bureau
    of Mines, 2401 E Street, N.W., Room 9017, Washington, DC
    20241; (202) 634-1265

Ronald M. Eng, Staff Officer, BRAB/NRC, 2101 Constitution
    Avenue, N.W., Washington, DC 20007; (201) 389-6542

E.K. Fields, Senior Research Associate, Amoco Chemicals
    Corporation, P.O. Box 400, Naperville, IL 60540; (312)
    420-5433

Robert E. Fisher, Manager of Planning, Cities Service
    Company, Box 300, Tulsa, OK 74102; (918) 586-2611

Arthur W. Francis, Product Manager, Union Carbide
    Corporation, UCC Technical Center, Old Saw Mill River
    Road, Tarrytown, NY 10591; (914) 345-3662

Bernard S. Friedman, Energy Consultant, 7321 South Shore
    Drive, Chicago, IL 60649; (312) 731-7886; (Chairman,
    Helium Conservation Task Force, American Chemical
    Society, Committee on Chemical and Public Affairs, DOE;
    (202) 376-4626)

Robert B. Fulton, Basic Materials Planning, Energy and
    Materials Department, E.I. duPont de Nemours and
    Company, Wilmington, DE 19898; (302) 774-5673

Richard D. Geiselman, General Manager, Marketing and Sales,
    Gardner Cryogenics, 2136 City Line Road, Bethlehem, PA
    18017; (215) 264-4523

R.C. Gilardi, Product Manager, Helium and Hydrogen, Airco
    Industrial Gases, 575 Mountain Avenue, Murray Hill, N.J.
    07974; (201) 464-8100, Ext. 271

Sharon Graham, Contract Specialist, National Science Foundation, 1800 G Street, N.W., Washington, DC 20550; (202) 632-5884

Ted Greenwood, Senior Policy Analyst, Office of Science and Technology Policy, Room 3026, New Executive Office Building, 17th and Pennsylvania Avenue, N.W., Washington, DC 20500; (202) 395-3265

William Gregory, Professor of Physics, Department of Physics, Georgetown University, Washington, DC 20057; (202) 625-4144

Robert W. Guernsey, Jr., Research Staff Member, IBM Research, Yorktown Heights, NY 10598; (914) 945-2639

Mary R. Hamilton, Manager, Energy Policy Department, The BDM Corporation, McLean, VA 22101; (703) 821-5304

Robert V. Hemm, Staff Engineer, National Materials Advisory Board, National Research Council, 2101 Constitution Avenue, N.W., Suite JH 426, Washington, DC 20418; (202) 389-6433

Mohamed A. Hilal, Assistant Professor, Department of Mechanical Engineering, University of Wisconsin, 541 ERB - 1500 Johnson Drive, Madison, WI 53706; (608) 263-2368

Charles W. Howe, University of Colorado, c/o Dept. of Agriculture and Applied Economics, 337 Classroom Office building, 1994 Buford Avenue, University of Minnesota, St. Paul, MN 55708; (612) 373-0951

H. Richard Howland, Senior Research Engineer, Westinghouse Research and Development Center, 1310 Beulah Road, Pittsburgh, PA 15235; (412) 256-3685

M. King Hubbert, Consultant, 5208 Westwood Drive, Washington, DC 20016; (202) 229-7798

Ralph P. Hudson, Chief, Heat Division, National Bureau of Standards, Room A311 - Physics Building, Washington, DC 20234; (202) 921-2034

John K. Hulm, Westinghouse Research and Development Center, 1310 Beulah Road, Pittsburgh, PA 15235; (412) 256-7222

Clyde Kimball, Program Director, Quantum Solids and Liquids, National Science Foundation, 1800 G Street, N.W. - DMR/QSL, Washington, DC 20550; (202) 632-7404

Clarence T. Kipps, Jr., Partner, Miller and Chevalier, 1700 Pennsylvania Avenue, N.W., Washington, DC 20006; (202) 393-5660

M.C. Krupka, Senior Staff Member, Los Alamos Scientific Laboratory, P.O. Box 1663, Los Alamos, NM 87544; (505) 664-5802

Lester B. Lave, Department of Economics, Carnegie-Mellon University, Pittsburgh, PA 15213; (412) 578-2290

Thomas S. Lincoln, Student, 361 Juniper Court, Herndon, VA 22070; (703) 437-5274

Franklin A. Long, Program on Science, Technology and Society, 632 Clark Hall, Cornell University, Ithaca, NY 14853; (607) 256-3810

Robert S. Long, Executive Secretary, Board on Mineral and Energy Resources, National Research Council, 2101 Constitution Avenue, N.W., Washington, DC 20418; (202) 389-6194

A.D. McInturff, Physicist, ISABELLE Division, Brookhaven National Laboratory, Upton, NY 11719; (516) 345-4845

Ray D. Munnerlyn, Chief, Division of Helium, U.S. Bureau of Mines, 2401 E Street, N.W., Washington, DC 20241; (202) 634-4734

F. Clayton Nicholson, Consultant, c/o Northern Natural Gas Company, 1133 - 15th Street, N.W., Washington, DC 20005; (202) 293-1260

Lewis H. Nosanow, Head, Condensed Matter Sciences Section, Division of Materials Research, National Science Foundation, 1800 G Street, N.W., Washington, DC 20550; (202) 632-7404

William L. Petrie (Project Officer), National Research Council, Suite JH 737, 2101 Constitution Avenue, N.W., Washington, DC 20418; (202) 389-6508

Ross David Pollack, Visiting Attorney, Environmental Law Institute, 150 Huntley drive, Ardsley, NY 10502; (914) 693-0724

J. Benjamin Reinoehl, Marketing Manager, Air Products and Chemicals, Inc., P.O. Box 11538, Allentown, PA 18105; (215) 398-8144

Richard L. Rerig, Marketing Manager, Air Products and
    Chemicals, Inc., 1919 Vultee Street, Allentown, PA
    18103; (215) 398-8331

J.C. Richards, Vice President, Pullman, Inc., 1616 H Street,
    N.W., Washington, DC 20006; (202) 638-5522

Nancy Robinson, Legislative Assistant to Congressman
    Sebelius (Dem.-Kansas), 1211 Longworth House Office
    Building, Washington, DC 20515; (202) 225-2715

Charles A. Romine, Industrial Economist, Midwest Research
    Institute, 425 Volker Boulevard, Kansas City, MO 64110;
    (816) 753-7600

Jay Russell, Staff Attorney, Environmental Law Institute,
    1346 Connecticut Avenue, N.W., Washington, DC 20036;
    (202) 452-9600

Gerald S. Schatz, Editor, News Report, National Academy of
    Sciences, 2101 Constitution Avenue, Washington, DC
    20418; (202) 389-6360

Armond Sonnek, Staff Assistant, U.S. Bureau of Mines, 2401 E
    Street, N.W., Washington, DC 20241; (202) 634-4734

V. Kerry Smith, Resources for the Future, 1755 Massachusetts
    Avenue, N.W., Washington, DC 20036; (202) 462-4400

Donald E. Stouffer, Supervisory Auditor, General Accounting
    Office, Room 6311W, 24th and E Streets, N.W.,
    Washington, DC 20548; (202) 254-7391

Michael Tinkham, Department of Physics, Harvard University,
    Cambridge, MA 02138; (617) 495-2866

Cynthia Tinberg, Economist, Midwest Research Institute, 425
    Volker Boulevard, Kansas City, MO 64110; (816) 753-7600

Paul A. Treado, Chairman, Department of Physics, Georgetown
    University, Washington, DC 20057; (202) 625-4144

Phil Tully, Helium Technologist, U.S. Bureau of Mines, Box
    H-4372 Herring Plaza, Amarillo, TX 79101; (806) 376-2604

Albert E. Utton, School of Law, 1117 Stanford, N.E.,
    University of New Mexico, Albuquerque, NM 87131; (505)
    277-4910

George C. Vaughan, Vice President, Natural Gas Liquids
    Marketing, Cities Service Company, Box 300, Tulsa, OK
    74102; (918) 586-2402

Irwin L. (Jack) White, Science and Public Policy Program,
601 Elm Avenue, University of Oklahoma, Norman, OK
73069; (405) 325-2554

Martin Zlotnick, Project Officer, Division of Power Systems,
Department of Energy, Washington, DC 20545; (202) 376-
4606

INVITATION AND RESPONSES

NATIONAL RESEARCH COUNCIL
COMMISSION ON NATURAL RESOURCES

2101 Constitution Avenue   Washington, D. C. 20418

BOARD ON MINERAL AND ENERGY RESOURCES

PUBLIC FORUM

HELIUM: PRESENT AND FUTURE NEEDS

November 20-21, 1977

The U.S. Bureau of Mines has asked the Board on Mineral and Energy Resources to evaluate the present federal program for the conservation of helium and its adequacy to meet future needs. A report of this study is to be submitted to the Bureau by January 1978.

To obtain the widest possible cross-sectional input of information, the study committee plans to convene a public forum in the National Academy of Sciences Auditorium, 2100 Block, C Street, N.W., Washington, D.C., at 6:00 p.m., Sunday, November 20, and resuming at 9:00 a.m., Monday, November 21, 1977, to discuss the following questions:

I.   Should helium be conserved in the national interest?

II.   Are the present stores of helium adequate?

III. Will more helium be needed in future years?

IV.   How should our national helium program be managed?

As the time for the preparation of this meeting is quite limited, we invite the submission of prepared statements (with abstracts) responsive to these questions to Mr. William L. Petrie, NAS-NRC, 2101 Constitution Avenue, N.W., Washington, D.C. 20418, by November 16. Although such advance material is sought by the committee as a basis for selecting speakers, it is not a prerequisite to attendance or participation in the discussions.

Please telephone Marcie Lofgren on (202) 389-6323 for further information and advance registration.

International Business Machines Corporation

November 17, 1977

Thomas J. Watson Research Center
P. O. Box 218
Yorktown Heights, New York 10598
914/945-2555

Mr. William L. Petrie
National Academy of Sciences
2101 Constitution Avenue
Washington, D.C. 20418

Dear Mr. Petrie:

It seems improbable that I shall able to be with the Helium Committee Tuesday November 22 in Washington. I have already indicated that it is impossible for me to attend the meeting at the Forum on November 20 or 21.

I do have two suggestions which might well find place in the report of the Committee. The first is simply to point out that there is a very considerable loss of value involved in the venting of the helium-rich gas which is currently produced when natural gas liquids and nitrogen are stripped from natural gas before the latter is shipped to the Middle West. The cost of putting this "crude helium" into storage is minimal, as is the cost of keeping it in storage in a depleted natural gas field. What seems to be preventing this rational action is the dispute between the helium producers and the Government as to whether the Government should pay for the helium. Surely it is not beyond the powers of man (or woman) to devise a kind of escrow scheme, whereby the helium can be put into storage without prejudice to the rights of either party if there is further legislation. That is, the helium would belong to the producer and could be withdrawn by him. In the very near term (months) this problem should be resolved to give us more time to attack the more important and longer-term questions.

Ultimately, we will have to extract helium from natural gas formations which yield so little natural gas that it is not economical to extract the natural gas. Alternatively, we shall be dependent (in the long run) upon extraction of helium from air. In that era, there will of course be a tremendous premium on reducing the loss of helium wherever helium is used, on reducing the inventory of helium in superconducting transmission lines, and the like. Still, if we are counting now on a lower-cost method for extracting helium from the air than those which are now known (especially in the face of rising energy costs) as a justification for not separating and storing helium from natural gas, then it is economically desirable to spend a certain amount of money on research for discovering and perfecting such a process (or alternatively, discovering that such a process does not exist).

If we were recovering all the helium from the natural gas

being supplied to consumers, then it <u>might</u> be legitimate to wait for a few decades before conducting such research (if one felt that ten or twenty years of exploration would give us all the information there was to be obtained). But when we are faced with a near-term decision as to whether or not to separate and store, this research should be done now.

I do not say that no economic technique can be devised to extract helium from air. I do say that the known methods are too expensive in energy or in capital to be competitive with storage in the near term. This is the basis for the rather elliptic calculation in the final statement of the American Physical Society. Clearly, an expenditure of a few million dollars per year would be economically justifiable, but initially all reasonable ideas could be funded for much less.

Let me volunteer my good wishes for the work of this Committee, but my prejudice that what is needed also is some analysis by one or two people sort of full time over a period of months.

Sincerely yours,

Richard L. Garwin

RLG:mll:321.WP

L. C. SMITH COLLEGE OF ENGINEERING
Department of Chemical Engineering and Materials Science

CHEMICAL ENGINEERING | 229 HINDS HALL
SYRACUSE, NEW YORK 13210
315/423-2557

November 16, 1977

Mr. William L. Petrie
National Research Council
Commission on Natural Resources
2101 Constitution Avenue, N.W.
Washington, D.C. 20418

Dear Mr. Petrie:

I have just noticed your announcement concerning the public forum
on the present and future needs for helium, to be held at the National
Academy of Sciences on November 20 - 21. Should it be decided that
continuing conservation of helium is in the national interest, I would
suggest that the Commission on Natural Resources evaluate the new method
of helium recovery from natural gas by selective permeation through
polymer membranes. This method is particularly well suited for producing a gas stream containing 70 - 80% helium for conservation purposes.
I will be happy to provide the necessary information, should there be any
interest in this matter.

Sincerely yours,

S. Alexander Stern
Professor

SAS/kd

HENRY M. JACKSON, WASH., CHAIRMAN

FRANK CHURCH, IDAHO
LEE METCALF, MONT.
J. BENNETT JOHNSTON, LA.
JAMES ABOUREZK, S. DAK.
FLOYD K. HASKELL, COLO.
DALE BUMPERS, ARK.
WENDELL H. FORD, KY.
JOHN A. DURKIN, N.H.
HOWARD M. METZENBAUM, OHIO
SPARK M. MATSUNAGA, HAWAII

CLIFFORD P. HANSEN, WYO.
MARK O. HATFIELD, OREG.
JAMES A. McCLURE, IDAHO
DEWEY F. BARTLETT, OKLA.
LOWELL P. WEICKER, JR., CONN.
PETE V. DOMENICI, N. MEX.
PAUL LAXALT, NEV.

GRENVILLE GARSIDE, STAFF DIRECTOR AND COUNSEL
DANIEL A. DREYFUS, DEPUTY STAFF DIRECTOR FOR LEGISLATION
D. MICHAEL HARVEY, CHIEF COUNSEL
W. O. CRAFT, JR., MINORITY COUNSEL

United States Senate

COMMITTEE ON
ENERGY AND NATURAL RESOURCES

WASHINGTON, D.C. 20510

November 15, 1977

Mr. William L. Petrie
NAS - NRC
2101 Constitution Avenue, N.W.
Washington, D.C. 20418

Dear Mr. Petrie:

In reply to your notice of a public forum on "Helium: Present and Future Needs," I would appreciate having this letter and its enclosures included in your record and made available to the Board on Mineral and Energy Resources.

With regard to the questions under consideration, I would submit the following replies:

    I.   It is definitely in the national interest to conserve helium.

   II.   The present stores of helium are <u>not</u> adequate for meeting future national needs.

  III.   More helium will certainly be required in future years.

  IV.   The national helium program should be managed in accordance with the 1960 Helium Act, as established by Congress, whereby the extraction and storage of helium would be reinstituted.

In support of my replies, I am enclosing a copy of my statement before the House Interior and Insular Affairs Committee on September 20, 1977, concerning helium conservation, and a copy of the Summary Report, "Helium – Its Storage and Use in Future Years." I hope that these will be useful to the Board and I look forward to the report of your study.

Sincerely,

James A. McClure
United States Senator

McC:fe

STATEMENT OF HONORABLE JAMES A. McCLURE,
A UNITED STATES SENATOR FROM THE STATE OF IDAHO

BEFORE THE SUBCOMMITTEE ON MINES AND MINING, HOUSE INTERIOR
AND INSULAR AFFAIRS COMMITTEE, CONCERNING
H.RES. 91 AND HELIUM CONSERVATION

TUESDAY, SEPTEMBER 20, 1977

Mr. Chairman, I commend you and your fellow Committee members on holding this hearing, concerning one of the basic questions facing our Nation's research and development efforts towards improved energy sources; namely, where will we get the supply of helium necessary to utilize the technology under development? Both the Administration and the Congress should pay close attention to your proceedings, with specific emphasis on the implications concerning the enormous expenditures being planned for Federal R&D where no practical use can be made of the results without helium. We have one department of Government spending huge sums for research and another department destroying the ability to use the results.

A brief summary of how we arrived at this point would be helpful. As you know, the debate over helium is certainly not a new one. Understanding the roots of this question can, I believe, help us to better evaluate our present situation and make the necessary decisions regarding the Federal role in helium production, conservation, and use.

To those not familiar with the helium issue, it would be worthwhile to remember that the original call for conserving helium came in 1957 from the Office of Defense Mobilization.

From this early warning evolved the decision of President Eisenhower to declare a National Helium Conservation Policy in April of 1958. It was this action by the Administration that resulted in this Committee's developing, with great care and foresight, the legislative background resulting in approval for the conservation of helium under the Helium Act Amendments of 1960. We are not faced with a sudden crisis, but with a problem which was recognized almost two decades ago and was adequately faced initially. In recent years, it has emerged as a serious problem due to unilateral action of the past Administration. I am referring, of course, to the decision of OMB to cancel the contracts under which helium was being extracted from natural gas and saved for future use. This decision was implemented by the Secretary of the Interior and is now being tested in the Courts.

My view on the legal questions involved is based on the elementary damage principle that in a breach of contract case the injured party is normally entitled to recover his profit plus his unavoidable cost of performance. Here, three of the contractors cannot avoid extracting the helium because they integrated the helium plants with other facilities. This integration was contemplated and encouraged by the Government. In his comprehensive Report to the U. S. Court of Claims in the Northern Helex case, the Trial Judge decided that the contractor was entitled to recover the full contract price and recommended judgment for $78 million.

In an extremely confusing split decision dated
October 22, 1975, the Court returned the case to the Trial
Judge. The Court held that the contractor was entitled to
be placed in the same monetary position it would have been in
had the Government performed rather than breached the contract, but said that the contractor was not entitled to recover its unavoidable cost of performance. How the incon-
sistencies in the October 1975 decision are to be resolved
remain for further decision by the Court. However, I believe
that when these cases are finally decided, we will find that
the U. S. taxpayer is going to pay the full contract price for
every cubic foot of helium which the Department of the Interior
is now venting to the atmosphere. In other words, the American
taxpayers will not only lose the benefits of conserving the
helium, they will also probably have to pay the costs which
would have been incurred under the contracts with the Department of the Interior and also pay the costs of extracting the
helium from the atmosphere at some time in the future. I have
repeatedly urged the

Administration to reconsider this ill-conceived termination. If they are not willing to reinstitute the conservation program, they should at least stop venting helium into the air until the final Court decision is in. Unfortunately, my position was not accepted and results in wastage of both money and helium.

The Energy Reorganization Act of 1974 included a direction to the Administrator of ERDA to conduct a study of the potential energy applications of helium. This study was completed and submitted by the Administrator. I believe that Dr. Hammel and his Staff at Los Alamos who did the work should be congratulated on the fine technical work that has resulted. While not being completely pleased with the conclusions and recommendations, I do believe that this work adds to our knowledge of the uses and needs for helium and will prove its usefulness in the coming years. The ERDA Report was reviewed by the Subcommittee on Energy Research & Development of the Science Committee on May 7, 1975.

As the ERDA Report so accurately states, the economic costs of recovering helium from the atmosphere will range between $3,000 and $6,000 per thousand cubic feet. This is, indeed, 150 to 300 times greater than the present commercial cost of helium and 350 to 700 times the cost of Bureau of Mines helium produced from existing plants. In addition, we must consider the impact of the energy required for the extraction process, which would be the equivalent of 70% of the projected annual output of the Alaskan oil pipeline, based on a helium output of one billion cubic feet per year. In other words, the extraction of 20 billion cubic feet of helium from the atmosphere (the amount being lost from existing extraction plants) would require the equivalent of about 10 billion barrels of oil.

ERDA, DOD, Commerce, the Electric Power Research Institute and other organizations have initiated national projects which expand the need for helium. These include airborne superconducting machines; marine and submarine electric drives to extend the performance, versatility and cruising range of ships; harnessing of thermonuclear energy to extend our energy sources; enhancement of power station efficiencies; systems for transmitting and storing electrical energy which reduce demands on land and basic resources while conserving energy; new communication techniques and many other advances in science and technology.

As we can now see, the population beyond 1990 will no doubt live in a world of increased complexity, greater instability and greater dependence on a wider range of technologies. There will be many more people, more pressure to utilize energy, land, water and material resources more efficiently. They will need all the options they can get. The continued availability of helium at low-energy cost and of a wide range of technologies involving superconductivity will enable them to have additional options for improving the efficiency of energy conversion and utilization, conserving resources, and maintaining survival options in a world of increasingly technical sophistication.

Space is a new and growing frontier with implications for peaceful and military uses. Both space and defense programs require the most advanced communications and computation systems. Global navigation systems incorporating satellite-based super-conducting gyroscopes are needed. Solid state Maser devices cooled to liquid helium temperatures offer the ultimate in microwave receiver sensitivity. The highest power gas Lasers use helium-gas mixtures. They offer the ultimate in ranging and communication over astronomical distances and permit the highest concentrations of light energy ever known, capable of vaporizing metal and exploding air over long distances. Infra red Lasers permit ranging and gunnery at night.

Prototype electric drives using superconducting motor-generator sets in naval ships have been constructed and success-fully tested. More powerful units in conjunction with nuclear energy sources are contemplated for use in large ships and tankers because of their increased flexibility, higher effi-ciencies and the possibilities these bring for extended cruis-ing range and increased speeds. Optional placement of the units will improve and simplify ship design.

It is when we examine ERDA's findings on which energy-related technologies will require large amounts of helium that we can fully appreciate the ultimate impact of making the wrong policy decision. Three of the areas specifically cited are

superconducting power transmission, superconducting magnetic
energy storage, and fusion reactors. I know that this Committee needs no testimony on the importance of these technologies for meeting our future energy needs. It is the risk of
foreclosing possible breakthroughs in these fields which makes
the question of helium so critical. I believe that the President's Energy R&D Advisory Council has accurately summarized
this critical nature in their resolution   regarding helium conservation, and I quote:

> "The greater risk attends a failure to
> anticipate the real possibility that the
> U. S. technological base in the 21st
> century is heavily dependent on large
> quantities of helium available at low
> energy cost."

It is this risk that we cannot afford to take.

Again I commend you for your time and effort on this
question. I look forward to working with you and other concerned colleagues in reestablishing a forward-looking national
helium policy. In the meantime, I believe H. Res. 91 (which is
essentially the same as S. Res. 253 passed by the Senate on
May 13, 1976) should be reported out by this Subcommittee and
passed by the House.

Thank you.

CITIES SERVICE COMPANY
BOX 300
TULSA, OKLAHOMA 74102

November 15, 1977

National Academy of Sciences
National Research Council
2101 Constitution Avenue, N.W.
Washington, D.C. 20418

Attention: Mr. William L. Petrie

Subject: Helium - Present and Future Needs

Dear Mr. Petrie:

   Thank you for an invitation to attend a public forum on helium. I understand it is sponsored by the National Academy of Sciences and will be held in Washington, D.C. on November 20 and 21, 1977.

   Cities Service Company, a leading producer and marketer of helium is vitally interested in the future preservation of this depleting natural resource. Other company representatives and I will be in attendance. I plan to present remarks on behalf of Cities Service, during the informal discussion period. Attached is an abstract and copy of my prepared statements.

   Please feel free to contact me by telephone if you need additional information. (Area Code 918, 586-2402)

       Yours very truly,

       NATURAL GAS LIQUIDS DIVISION

       George C. Vaughan, Vice President
       Marketing and Supply & Distribution

HELIUM: PRESENT AND FUTURE NEEDS
ABSTRACT OF COMMENTS BY CITIES SERVICE HELEX, INC.,
BEFORE THE NATIONAL RESEARCH COUNCIL

I.   Should helium be conserved in the national interest?

Increasing reliance on foreign sources for conventional forms
of energy poses a substantial threat to U. S. security. Since
major proved reserves of helium are commingled in declining
natural gas streams, a natural resource which may be needed
for development of future domestic energy sources, could be
lost forever.

II. Are present stores of helium adequate to meet future needs?
(Incorporates the National Research Council's questions II
and III)

In 1976, the U. S. Senate passed a Resolution acknowledging
the potential requirements for helium in development of future
domestic energy resources. This action was supported by a
detailed report prepared by the Energy Research and Development Administration. In short, positive action must be exerted
to preserve proven helium reserves for use at a possible time
of critical need.

III. How should our national helium program be managed?

The Federal Government must accept a dominant role through
legislation conducive to a national helium conservation program.
Answers to issues covering production, lease, tax and storage
liabilities must be mutually acceptable to government and
private industry.

HELIUM: PRESENT AND FUTURE NEEDS
STATEMENT OF CITIES SERVICE HELEX, INC.,
BEFORE THE NATIONAL RESEARCH COUNCIL

I. Should helium be conserved in the national interest?

The national interest can best be served through conserving
helium, a unique, valuable but disappearing natural resource.
Our country's primary helium proved reserves are contained
in declining natural gas streams. Because there is no known
substitute for present applications, including related technological research, future helium supplies may be jeopardized as
natural gas fields are depleted to meet consumer fuel
requirements.

Today, U.S. reliance on foreign oil increasingly presents a
threat to our national well-being. As we replace conventional
fuels with other forms of energy, such as coal, nuclear and
solar, strategic helium reserves may well serve to help preserve our national economy and security for future generations,
while providing a large contribution to efficient use of energy.

II. Are present stores of helium adequate to meet future needs?

On April 11, 1975, the Energy Research and Development
Administration, presented an extensive study entitled "The
Energy Related Applications of Helium". The report

essentially states that energy-related demands for helium, (superconducting power transmission, superconducting magnetic energy storage and fusion reactors) could start around the year 2000, increasing to 1-5 billion cubic feet/year by 2030. Consequently, known technology to preserve helium may be wasted and today's reserves depleted at a time of critical need. This was also recognized by a significant majority of the U. S. Senate, which passed Resolution S. Res. 253, on May 13, 1976. But to date, no program has been implemented to conserve excess helium production.

Future planning must also address the increasing current demands for helium. These include numerous vital industries such as underway operations associated with offshore oil exploration and production, medical-- hospital and research, heli-arc welding, protective atmosphere and leak detection and chromatography.

Emphasis must be directed toward storing presently produced helium in excess of market demands, while also providing incentive to reactivate and retain production from idle production facilities. Although

helium exists in the atmosphere, this does not

represent a viable economic option when compared

to existing technology. Furthermore, extraction of

helium from low concentrations in air would require

tremendous quantities of energy at a time of fore-

casted energy shortage.

III. How should our national helium program be managed?

Leadership from the federal government is a

necessary prerequisite to re-establishing a national

helium conservation program. All existing storage

facilities are owned by the government which are

capable of retaining this depleting resource. Private

companies can not financially absorb production costs

well into the future. Answers to other monetary

problems, including contract and lease payment liti-

gations, taxes and storage costs must also be develop-

ed. Accordingly, government, industry and the -

scientific community must work collectively to provide

solutions and new legislation that will preserve helium

for the future material needs of our society.

BROOKHAVEN NATIONAL LABORATORY

ASSOCIATED UNIVERSITIES, INC.

Upton, New York 11973

Accelerator Department

(516) 345- 3454

November 14, 1977

Mr. William L Petrie
NAS — National Research Council
2101 Constitution Avenue, NW
Washington, D.C.   20418

        Subject: Your memo regarding helium conservation

Dear Sir:

Today I received your note concerning the public forum on helium
needs.

As you mentioned, time is short to give comments; in any case mine
are strictly qualitative based on working in the cryogenic field
with superconducting devices.

I feel very strongly about two questions which you posed:

   I.   Should helium be conserved in the national interest?

        In reply, since there is a limited amount of helium
        available and since it is a by-product in processing
        natural gas, which is now being thrown away, the answer
        to this question should be a positive "YES".

   III. Will more helium be needed in future years?

        The technology of superconducting devices has expanded
        tremendously in the last ten years, and the amount of
        helium required has seen an equal growth. Based on
        present growth the future needs will probably increase
        at an even greater rate so that, in my opinion, the helium
        needs will greatly increase in future years.

                              Sincerely yours,

                              Joseph A. Bamberger

JAB/ar
cc: D.P. Brown

General Motors Research Laboratories
Warren, Michigan 48090

November 11, 1977

Mr. William L. Petrie
NAS-NRC
2101 Constitution Ave. N.W.
Washington, D.C. 20418

Dear Mr. Petrie:

Thank you for bringing to our attention the public forum on the present and future needs of helium. It is important that the United States provide for the future needs of this unique gas. General Motors uses helium in a few of its welding, sputtering and plasma spraying procedures, for gas leak detection, and in our research for new materials and processes. It would also be an important material for any future technology which might be developed around the use of superconductivity.

Helium supplies a non-reactive atmosphere for arc welding or sputtering of highly reactive materials such as aluminum. It is singular in its ability to penetrate narrow cracks and pin holes and hence is valuable to locate potential gas leaks in systems such as refrigeration or air-conditioner compressors and condensers. It is essential in liquid form in research at temperatures near zero K where many fundamental properties of materials can be measured and compared to basic theory. And it is essential for obtaining superconducting materials which may result in new energy saving technology. However, there is considerable research to try to obtain materials with very high conductivity at temperatures which may be obtainable by using liquid hydrogen instead of liquid helium.

We cannot comment directly on the four questions being addressed by the public forum at present because we have not considered helium supply a critical problem. However, we are certain that we will continue to need a moderate supply of helium in future years for the applications referred to above and are pleased that the Bureau of Mines is addressing this problem.

Very truly yours,

David L. Fry

David L. Fry
Departmental Research Scientist
Physics Department

DLF/jw

cc: Joseph B. Bidwell
    John D. Caplan
    Nils L. Muench
    Frank E. Jamerson

VIRGINIA POLYTECHNIC INSTITUTE AND STATE UNIVERSITY

Blacksburg, Virginia 24061

UNIVERSITY DISTINGUISHED PROFESSOR OF ENGINEERING (703) 951-6473

November 8, 1977

Mr. William L. Petrie
NAS-NRC
2101 Constitution Avenue, NW
Washington, DC 20418

Dear Mr. Petrie:

I am responding to your notice of a Public Forum on Helium: Present and Future Needs on November 20-21, 1977. From 1965 - 1968, I was Director of the US Bureau of Mines and thus responsible for the Helium Conservation Program for those years.

It was the only truly mineral conservation program in existence in the federal government at that time. It suffered from several problems:

(1) Fixed price purchase contracts

(2) Industry competition at lower prices

(3) Government support of the industry competition

As a result it was terminated.

There is no way of forecasting the needs for helium in the future. However, the sources of relatively inexpensive helium are being used. Therefore, these sources are being irreversibly exhausted.

It seems to me, therefore, that a new helium conservation program is required to avoid helium waste; that it should be administered in such a way as to provide incentives for industry to recover helium or retain the helium bearing gases; and that it should be administered in such a way that there are no windfall profits or fall guys. It might be simplest for the government to purchase the reserves of helium bearing natural gas and hold them as a stockpile or reserve.

Sincerely

Walter Hibbard

Walter R. Hibbard
University Distinguished Professor of Eng.
Director-Va. Ctr. for Coal & Energy Research

WRH/bj

UNIVERSITY OF WISCONSIN-MADISON
THEORETICAL CHEMISTRY INSTITUTE
1101 UNIVERSITY AVENUE
MADISON, WISCONSIN 53706

PH O. HIRSCHFELDER
DIRECTOR

CURTISS
ASSOCIATE DIRECTOR

16 July 1976

TELEPHONE:
262-1511
AREA CODE 608

Professor Richard B. Bernstein
Department of Physics
University of Texas at Austin
Austin, Texas   78712

Dear Dick:

I am very much concerned with the future shortage of helium. I
would appreciate it very much if you would bring this matter to the
attention of the National Research Council. I have written to
Senator Proxmire and Senator McCormack in this regard and would very
much appreciate your help.

Large amounts of liquid helium will be required if the United States
is to develop controlled nuclear fusion as a source of energy; if the
United States is going to store energy in the form of low-temperature
·superconducting storage rings; or if the United States is going to
transmit high-voltage electrical energy in superconducting pipes. The
production of liquid helium from the government helium wells in Amarillo
was stopped because it was (erroneously) asserted that their operation
was not economical. Instead, the government encouraged the natural gas
companies to set up facilities for producing helium from natural gas.
I understand that these natural gas companies were promised that the
government would buy all of the liquid helium which they produced.
However, I have heard recently that the government is no longer buying
this liquid helium and the natural gas companies are venting their
helium. As you know, the world supply of helium is very limited, and
helium is a national resource which requires conservation. Probably,
it also requires a modest government subsidy to conserve it. I would
appreciate any help which you might be able to furnish. (See for
example, the 1974 RAAN Report of Charles Laverick entitled "Helium-Its
Storage in Future Years").

With very best wishes,

Cordially,

Joe

Joseph O. Hirschfelder

JOH:gl